Translational Bioinformatics

Volume 9

Series editor

Xiangdong Wang, MD, Ph.D.
Professor of Medicine, Zhongshan Hospital, Fudan University Medical School, China
Director of Shanghai Institute of Clinical Bioinformatics, (www.fuccb.org)
Professor of Clinical Bioinformatics, Lund University, Sweden

Aims and Scope

The Book Series in Translational Bioinformatics is a powerful and integrative resource for understanding and translating discoveries and advances of genomic, transcriptomic, proteomic and bioinformatic technologies into the study of human diseases. The Series represents leading global opinions on the translation of bioinformatics sciences into both the clinical setting and descriptions to medical informatics. It presents the critical evidence to further understand the molecular mechanisms underlying organ or cell dysfunctions in human diseases, the results of genomic, transcriptomic, proteomic and bioinformatic studies from human tissues dedicated to the discovery and validation of diagnostic and prognostic disease biomarkers, essential information on the identification and validation of novel drug targets and the application of tissue genomics, transcriptomics, proteomics and bioinformatics in drug efficacy and toxicity in clinical research.

The Book Series in Translational Bioinformatics focuses on outstanding articles/chapters presenting significant recent works in genomic, transcriptomic, proteomic and bioinformatic profiles related to human organ or cell dysfunctions and clinical findings. The Series includes bioinformatics-driven molecular and cellular disease mechanisms, the understanding of human diseases and the improvement of patient prognoses. Additionally, it provides practical and useful study insights into and protocols of design and methodology.

Series Description

Translational bioinformatics is defined as the development of storage-related, analytic, and interpretive methods to optimize the transformation of increasingly voluminous biomedical data, and genomic data in particular, into proactive, predictive, preventive, and participatory health. Translational bioinformatics includes research on the development of novel techniques for the integration of biological and clinical data and the evolution of clinical informatics methodology to encompass biological observations. The end product of translational bioinformatics is the newly found knowledge from these integrative efforts that can be disseminated to a variety of stakeholders including biomedical scientists, clinicians, and patients. Issues related to database management, administration, or policy will be coordinated through the clinical research informatics domain. Analytic, storage-related, and interpretive methods should be used to improve predictions, early diagnostics, severity monitoring, therapeutic effects, and the prognosis of human diseases.

Recently Published and Forthcoming Volumes

Single Cell Sequencing and Systems Immunology
Editors: Xiangdong Wang, Xiaoming Chen, Zhihong Sun, Jinglin Xia
Volume 5

Computational and Statistical Epigenomics
Editor: Andrew E. Teschendorff
Volume 7

Genomics and Proteomics for Clinical Discovery and Development
Editor: György Marko-Varga
Volume 6

Allergy Bioinformatics
Editors: Ailin Tao, Eyal Raz
Volume 8

More information about this series at http://www.springer.com/series/11057

Jiaqian Wu
Editor

Transcriptomics and Gene Regulation

Honor editor
Dong Kim

Springer

Editor
Jiaqian Wu
The Vivian L. Smith Department of Neurosurgery
Center for Stem Cell and Regenerative Medicine
The University of Texas Medical School at Houston
Houston, TX
USA

ISSN 2213-2775 ISSN 2213-2783 (electronic)
Translational Bioinformatics
ISBN 978-94-017-7448-2 ISBN 978-94-017-7450-5 (eBook)
DOI 10.1007/978-94-017-7450-5

Library of Congress Control Number: 2015952061

Springer Dordrecht Heidelberg New York London

Printed on acid-free paper

Springer Science+Business Media B.V. Dordrecht is part of Springer Science+Business Media (www.springer.com)

Preface

This volume focuses on the modern computational and statistical tools for gene expression and regulation research to improve the understanding prognosis, diagnostics, prediction of severity, and therapies for human diseases. The recent advancements of microarray and next-generation sequencing technologies made it possible to detect gene expression at a genome-wide scale, which has greatly facilitated the identification of pathophysiological changes in various diseases. How are the global gene expression profiles regulated and what are the mechanisms underlying these changes? How are the signaling pathways altered under pathological conditions? How do the various regulatory molecules interact in a network to control disease states? How does the different genetic makeup of individuals affect the disease perceptibility and treatment outcome? These are fundamental questions for finding cures for complex human diseases and developing personalized medicine that is the future of health care. In this volume, we introduce the readers to some of state-of-the-art technologies as well as computational and statistical tools for translational bioinformatics in the areas of gene transcription and regulation, including the tools for next-generation sequencing analyses, alternative spicing, the modeling of signaling pathways, network analyses in predicting disease genes, as well as protein and gene expression data integration in complex human diseases. This volume is particularly suitable for researchers, physicians, or students in the field of molecular, clinical biology and bioinformatics. This exciting volume would not be possible without the expertise and dedication of all the contributing authors. Finally, I would like to dedicate this volume to my family for their unconditional love and support.

Houston, TX Jiaqian Wu

Contents

1 **The Analyses of Global Gene Expression and Transcription Factor Regulation** 1
Raquel Cuevas Diaz Duran, Sudheer Menon and Jiaqian Wu

2 **Global Approaches to Alternative Splicing and Its Regulation—Recent Advances and Open Questions** 37
Yun-Hua Esther Hsiao, Ashley A. Cass, Jae Hoon Bahn, Xianzhi Lin and Xinshu Xiao

3 **Long Noncoding RNAs: Critical Regulators for Cell Lineage Commitment in the Central Nervous System** 73
Xiaomin Dong, Naveen Reddy Muppani and Jiaqian Wu

4 **Gene Expression Models of Signaling Pathways** 99
Jeffrey T. Chang

5 **From Gene Expression to Disease Phenotypes: Network-Based Approaches to Study Complex Human Diseases** 115
Quanwei Zhang, Wen Zhang, Rubén Nogales-Cadenas, Jhin-Rong Lin, Ying Cai and Zhengdong D. Zhang

6 **Introduction to Isoform Sequencing Using Pacific Biosciences Technology (Iso-Seq)** 141
Manuel L. Gonzalez-Garay

7 **Systematic and Integrative Analysis of Gene Expression to Identify Feature Genes Underlying Human Diseases** 161
Zixing Wang, Wenlong Xu and Yin Liu

About the Editors

Prof. Jiaqian Wu An assistant professor in the Vivian L. Smith Department of Neurosurgery and Center for Stem Cell and Regenerative Medicine at the University of Texas Medical School at Houston, Dr. Wu earned her doctorate in molecular and human genetics at Baylor College of Medicine and did her postdoctoral work at Yale and Stanford University. The Wu laboratory combines stem cell biology and systems-based approaches involving functional genomics, bioinformatics, and next-generation sequencing technologies to unravel gene transcription and regulatory mechanisms governing neural and blood development and differentiation. Dr. Wu's work has been recognized with prestigious honors and awards, including the National Institute of Health Pathway to Independence (PI) Award (K99/R00), R01 and the Senator Lloyd and B.A. Bentsen Investigator Award which she currently holds; the National Institutes of Health Ruth L. Kirschstein National Research Service Award for Individual Postdoctoral Fellows; and the International Society for Stem Cell Research (ISSCR) Annual Meeting Travel Award. A reviewer for NIH, MRC, the journals Nucleic Acids Research, Genome Research, and Genome Biology, Dr. Wu has presented invited talks and lectures on stem cell biology, functional genomics, and proteomics at international conferences, the Multiple Sclerosis Research Center of New York, Lawrence Livermore National Laboratory, and the University of Florida, etc. She has developed a patent, authored a book, and wrote many articles that have appeared in PNAS, the Journal of Neuroscience, Genome Biology, Plos Genetics, Genome Research, and Nature, among others.

Prof. Dong H. Kim Dr. Kim is Professor and Chair of the Department of Neurosurgery at the University of Texas in Houston. As director of the Mischer Neuroscience Institute (MNI), he also leads the clinical neuroscience efforts for the Memorial Hermann Healthcare System. Currently, his group includes over 100 faculty and residents/fellows. A graduate of Stanford and the University of California, San Francisco (UCSF) School of Medicine, he completed general surgery training at Harvard, then neurosurgery at UCSF. Prior to coming to Texas, he held positions at Harvard Medical School, Brigham and Women's Hospital, the Dana-Farber Cancer Institute, Cornell University Medical College, The New York Hospital and Memorial Sloan Kettering Cancer Center. Dr. Kim's research has focused on the origin, development, and treatment of brain aneurysms. His group recently identified the first gene defect proven to cause intracranial aneurysms in familial patients. His research effort is also on stem cell therapy for treating spinal cord and brain injury. Dr. Kim was mentioned in the US News and World Report's Top 1% Doctors, and America's Top Surgeons. He is the recipient of grants from the National Institutes of Health and the American Stroke Association.

Chapter 1
The Analyses of Global Gene Expression and Transcription Factor Regulation

Raquel Cuevas Diaz Duran, Sudheer Menon and Jiaqian Wu

Abstract A major challenge in molecular cell biology lies in understanding how the same genome can give rise to different cell types and how gene expression is regulated. Gene expression and regulation studies focus on the abundance and structure of transcripts as well as how RNA production is controlled. High-throughput sequencing technologies such as RNA sequencing have allowed more accurate profiling of the transcriptome and the rapid identification of differentially expressed genes among samples. The regulation of gene expression is orchestrated by transcription factors. The development of ChIP sequencing assay has made it possible to comprehensively identify transcription factor-binding sites in vivo, allowing rapid unraveling of signaling pathways. The following chapter described the common methods used in studying global gene expression and transcription factor regulation with a special emphasis on bioinformatic analyses. The final section illustrates an example of an integrated gene expression and regulation study for identifying key factors regulating self-renewal and differentiation in hematopoietic precursor cells.

Keywords RNA sequencing · ChIP sequencing · Transcription factors · Transcriptome

R. Cuevas Diaz Duran · S. Menon · J. Wu (✉)
The Vivian L. Smith Department of Neurosurgery, University of Texas Medical School at Houston, Houston, TX 77030, USA
e-mail: jiaqian.wu@uth.tmc.edu

R. Cuevas Diaz Duran · S. Menon · J. Wu
Center for Stem Cell and Regenerative Medicine, UT Brown Institution of Molecular Medicine, Houston, TX 77030, USA

J. Wu, *Transcriptomics and Gene Regulation*,
Translational Bioinformatics 9, DOI 10.1007/978-94-017-7450-5_1

1.1 Introduction

Gene transcription and regulation are important areas of study because they underlie many biological processes and phenotypic variations in living organisms. Aberrant gene expression and regulation lead to diseases. The transcriptome consists of all transcripts synthesized in an organism including protein-coding, noncoding, alternatively spliced, polymorphic, sense, antisense, and edited RNAs. Transcriptome data analyses, namely the analyses of gene expression levels and structures, are essential for interpreting the functional elements of the genome and understanding the molecular constituents of cells and tissues. The regulation of gene expression is a basic mechanism through which RNA production is coordinated, and it controls important events such as development, homeostasis, and responses to environmental stimuli. Transcription factors, a type of DNA-binding proteins which recognize specific sequences, and other proteins work together through a variety of mechanisms to regulate gene transcription.

In this volume, different aspects regarding the analyses of gene transcription and regulation are described in individual chapters. In this chapter, we focus on gene expression level analyses by RNA sequencing (RNA-Seq) and transcription factor regulation by chromatin immunoprecipitation coupled with sequencing (ChIP-Seq). First, we review some useful methods developed in the past for characterizing global gene transcription.

1.2 Methods for Characterizing Global Gene Transcription

1.2.1 EST Sequencing

It is widely recognized that expressed sequence tag (EST) sequencing has provided an invaluable resource for identification of novel human genes [1, 2]. EST clustering methods allow EST to be systematically mapped, so that the information is readily integrated into the positional cloning project UniGene database (http://www.ncbi.nlm.nih.gov/UniGene). Because ESTs are from single-pass sequencing, they have to be carefully analyzed to remove genomic DNA and other contaminating sequences, such as mitochondrial, ribosomal, vector, and bacterial sequences. However, EST databases still contain a significant portion of (estimated 5–10 %) artifact sequences such as intronic or intergenic DNA [3, 4]. This is likely due to the presence of heterogeneous nuclear RNA (hnRNA) in RNA samples where EST libraries were generated. Moreover, it is difficult to understand the relationships among short EST sequences. EST clustering may confuse genes sharing similarities and alternatively spliced transcripts. Additionally, because of their short length and generally low quality, ESTs only provide limited information about gene structure and function [1]. Since EST sequencing is biased toward genes with high expression levels, the transcripts which are tissue-specific or low-abundant are less likely to be disclosed by EST sequencing. Therefore, methods that are less biased, more accurate, and sensitive are needed [5].

1.2.2 SAGE

The serial analysis of gene expression, or SAGE [6], is another technique used to quantify gene expression levels. SAGE method is designed to add a 9–14 bp tag adjacent to an *Nla*III restriction site at the gene's 3′ end. It measures transcript levels by automatically sequencing and counting each SAGE tag. The expression level of SAGE tags is analyzed and accessioned through the GEO repository. Additionally, an anatomic viewer "SAGE Genie" makes it easy to search and visualize the transcription level in different tissues and cell types of the human body (http://cgap.nci.nih.gov/SAGE). However, SAGE is a high-throughput technology which measures not the expression level of a gene, but a "tag" that represents a transcript. Due to alternative transcription, sequencing errors, and other potential effectors, sometimes two or more genes share the same tag or one gene may have more than one tag. Thus, the potential loss of fidelity should be taken into consideration.

Long serial analysis of gene expression (LongSAGE) is an adaptation of the original SAGE approach that can be used to rapidly identify novel genes and exons [7]. Instead of using an *Nla*III restriction site, LongSAGE uses a modification of longer tags (21 bp) added to a different restriction site (*Mme*I). The 21 bp tags include a constant 4 bp restriction site sequence where the transcript was cleaved and a unique 17 bp adjacent sequence of each transcript. The advantage of LongSAGE is the uniqueness of each tag in the human genome, which is not guaranteed by 14 bp tags. Sequencing tag concatamers and searching for the localization of tags in the genome help to verify predicted genes and to identity novel transcripts. LongSAGE has been reported to be at least an order of magnitude more efficient than EST sequencing [8].

1.2.3 Full-Length CDNA Sequencing Projects

In order to better access the biological information of genes including location of open reading frames, 5′ and 3′ untranslated regions, and splicing patterns, full-length and high-quality sequences of cDNAs are needed. cDNA sequencing is a valuable resource not only for characterizing the structure and function of known genes, but also for discovering novel genes. Especially with the completion of the human genome, the comparisons of the full-length and high-quality cDNA sequences with genomes are especially useful in identifying alternative gene structure and better understanding transcriptome composition during physiological and disease processes. Moreover, full-length cDNA sequencing projects paved the way for proteomic study by identifying new enzymes and proteins, generating physical clones for expression profiling, testing protein interactions, and generating hypotheses for biochemical studies.

There have been multiple efforts that aim at capturing the sequence of full-length clones which can be directly obtained from cDNA libraries generated from mammals and other selected organisms, such as *zebra fish, drosophila,* and *Caenorhabditis elegans*, mouse, and human [9–15]. In particular, NIH Mammalian Gene Collection project, utilizing large-scale RT-PCR-based cloning methods, provided thousands of clones of full-length human and mouse open reading frames [16–18].

1.2.4 Microarrays

DNA microarray is a hybridization-based technology which enables researchers to analyze the expression of large number of genes in a single reaction. DNA microarray technology was developed in the early 1990s. The technical advancement of this methodology is to manufacture slides or chips with thousands of nucleic acid probes immobilized on a small surface area. DNA probes are conformed of several specific DNA sequences of genes to which cDNA samples are hybridized. Samples, also referred to as targets, may be obtained from cells in different biological or experimental conditions, tissues, organisms, or developmental stages. Probe–target hybridization is quantified by fluorescent labeling. The signal intensities captured as images after scanning are converted into a data matrix and processed using software specific to the application of the array to determine relative abundance of specific cDNA sequences from the samples. The DNA microarray is an effective tool to investigate the structure and activity of genes at a genome-wide scale, and it helps to elucidate the molecular mechanisms underlying normal and dysfunctional biological processes [19–24].

1.2.5 RNA-Seq

Even though microarray technology has provided valuable insights into gene function throughout the last decade, it suffers from limitations in resolution, dynamic range, and accuracy. The recently developed RNA-Seq methodology uses next-generation sequencing (NGS) to sequence cDNAs generated from RNA samples producing millions of short reads. The number of reads mapped within a genomic feature of interest (such as a gene or an exon) can be used as a measurement of the feature's abundance in the analyzed sample. Typical RNA-Seq procedure is depicted in Fig. 1.1. Briefly, RNAs are converted to a library of cDNA fragments with adaptors attached to one or both ends. The molecules, with or without amplification, are then sequenced with high-throughput technology, and

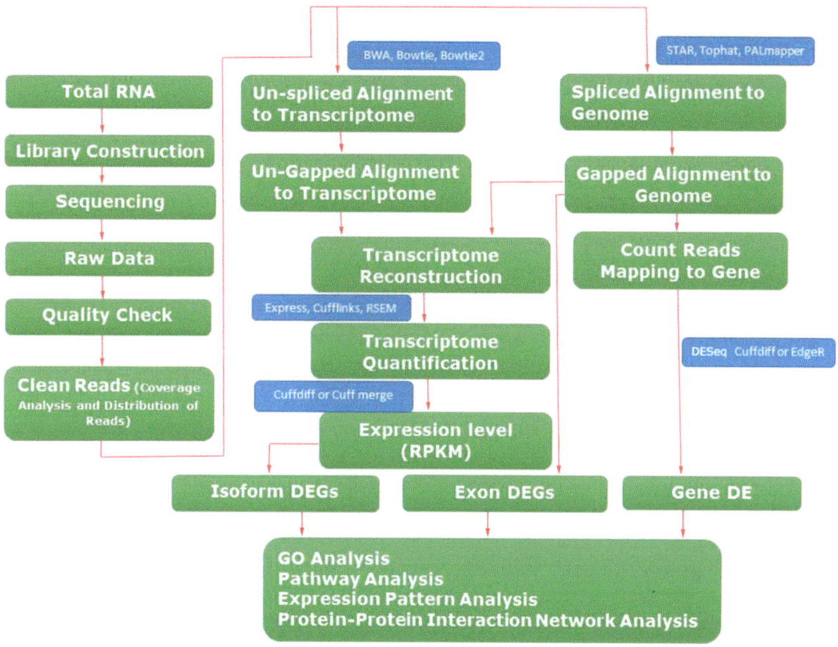

Fig. 1.1 RNA-Seq workflow. RNA-Seq begins with isolation of high-quality total RNA followed by conversion into cDNA, fragmentation, and adaptor ligation. Fragmented cDNA is used to construct a library for sequencing. Raw data, consisting of reads of a defined length, are preprocessed according to a set of quality control metrics, such as base quality, minimum read length, untrimmed adaptors, and sequence contamination. After filtering and trimming, reads are aligned to a reference genome or transcriptome depending on the objective of the experiment and the nature of the samples. Subsequently aligned reads are assembled. RNA-Seq assembly involves merging reads into larger contiguous sequences based on similarity. After assembly, reads are quantified in order to measure transcriptional activity. Read counts are generally computed in RPKM of FPKM in order to perform further downstream analysis, such as differential expression, pathway and gene set overrepresentation analysis, and interaction networks

short sequences from one end or both ends are obtained. The reads are typically 30–400 bp, depending on the DNA sequencing technology used. There are various high-throughput sequencing platforms available for RNA-Seq such as Illumina, Applied Biosystems SOLiD, and Roche 454 Life Science systems. The total reads obtained after sequencing are either aligned to a reference genome or transcriptome, or assembled de novo without genomic sequence guidance to create a genome-scale transcription map providing both transcriptional structure and expression level for each gene.

RNA-Seq has several advantages over microarrays [25–27]. First, sequencing technology is much more sensitive and quantitative than microarrays and it can provide a large dynamic range of detection (>9000-fold) [28]. Additionally, sequencing data are more specific and have less background. Moreover, sequencing experiments do not depend on the limited features of tiled microarrays and can therefore be used to interrogate any location in the genome and to detect and quantify the expression of previously unknown transcripts and splicing isoforms. Finally, sequencing is not limited by array hybridization chemistry, such as melting temperature, cross-hybridization, and secondary structure concerns.

1.2.5.1 Data Analysis General Workflow

RNA-Seq experiments result in millions of short sequence reads which require computational methods for comprehensive transcriptome characterization and quantification. Steps for data analysis vary according to the desired biological problem to be assessed and to the availability of reference genome or transcriptome data. A generic overview of the routine analyses performed is included in Fig. 1.1. The main tasks of data analysis are read mapping also referred to as alignment, transcriptome assembly, expression quantification, and downstream applications. Steps for data analysis, although it is sequential, may be performed with different computational tools and algorithms which require specific data formats and external files. It is desirable to automate the multiple data analysis steps in a pipeline. A pipeline is a reusable script with defined inputs, outputs, and parameters for each processing step. Several software platforms which collect different RNA-Seq analysis tools for each task have been developed such as PRADA [29], Tuxedo [30], MAP-R-SEQ [31], and GENE-Counter [32]. However, pipelines may be custom-built by users, selecting the most appropriate tools according to experimental data and desired downstream analysis [33]. The following subsections describe the different RNA-Seq data analysis tasks.

1.2.5.2 Quality Control and Preprocessing

The first step in data analysis is quality control. Accuracy in library preparation and sequencing steps contribute to the quality of reads, which if overseen may lead to erroneous mappings, misassembles, and false expression estimates. The quality control should include the assessment of read length, GC content, sequence complexity, sequence duplication, polymerase chain reaction artifacts, untrimmed adapters, low-confidence bases, 3′/5′ positional bias, sequence contamination, and fragment biases [34]. Quality control metrics are obtained directly from raw reads. Raw reads are typically in FASTQ format, text-based files which contain a sequence identifier, a nucleotide sequence, and its corresponding quality score [35]. A brief overview of the most important quality control processes to be performed is described next.

(a) Base Quality: Since RNA-Seq technologies rely on complex interactions between chemistry, hardware, and optical sensors, sequencing platforms provide metrics for assessing error probabilities. Base quality is measured by computing the confidence on base calling, the process by which the sequencer analyzes colorimetric sensor signals to predict individual bases. Base calling quality values are expressed in *Phred* scale, an error probability log transformation which has the advantage of converting low error probabilities to high-quality values and vice versa [36]. Quality values q are calculated for each base as:

$$q = -10\log_{10}(p) \tag{4.1}$$

where p is the estimated error probability of a base call. Base quality values are encoded with ASCII characters with the sequence data in FASTQ format [35]. Typically, good reads should have base qualities greater than 20; however, this threshold depends on the platform used. It is important to inspect the reads' base quality distribution to detect regions of poor base quality, which may be filtered or trimmed preserving the order of reads, thus increasing mapping efficiency. This process is also referred to as quality trimming and will be discussed next.

(b) Filtering and Trimming: Reads should be inspected for the presence of sequencing adapters, tags, and contaminating sequences, which should be removed before quality control processing. Adapter sequences and tags used during library construction should be removed prior to mapping. Contaminating sequences such as DNA, rRNA, or sequences from other organisms or vectors should be filtered out.

Additionally, reads should be filtered according to mean base quality or to the proportion of bases whose quality is below a user-defined threshold. There is no consensus on the optimal base quality threshold for trimming, and it is rather a trade-off between mapping efficiency and coverage. Software for filtering generally outputs synchronized filtered reads. When low-quality bases are located at the ends of reads, trimming is a better option than filtering. The basic principle of read trimming is to assess base quality keeping the longest possible high-quality read segments. Trimming with respect to base quality may be performed using running sum algorithms or sliding window-based algorithms. Running sum algorithms basically find the summation of the differences between all base quality values against a quality threshold, and sequences are trimmed at the base that makes the running sum minimum. When analyzing reads with a sliding window, the user defines a window size and a mean base quality threshold. Depending on the tool used, the window may slide from the 5′ or the 3′ ends of reads. Sliding the window from the 5′ end will trim the read until the window's quality lies below a threshold, maintaining the beginning of the sequence, whereas sliding from 3′ end will trim until a passing quality window is encountered. An excellent evaluation on the performance of several trimming tools was published by Del Fabbro et al. [37]. Some common tools used for trimming are Trimmomatic [38], Cutadapt [39], PRINSEQ [34], and ConDeTri [40].

Reads with a high frequency of ambiguous bases, bases not identified during sequencing, should be filtered out since they can lead to erroneous mapping and misassemblies. Low-complexity sequences (homopolymers, di-trinucleotides) will also result in mapping errors and should therefore be trimmed.

(c) GC Content Determination: Another important metric that should be considered is reads' mean GC content, which if plotted should follow a normal distribution centered on the organism's normal content. Variations on the GC content might be due to the PCR amplification process and therefore may be sample specific. An approach to reduce GC content systematic bias is conditional quantile normalization, a technique in which the distribution of read counts is modified by estimating quantiles obtained with a median regression on a subset of genes [41].

(d) Minimum Read Length: The distribution of read lengths should be verified after trimming since reads may have ended up as very short fragments, which become difficult for mapping. The minimum read length is a user-defined variable, and its value depends on desired downstream applications.

(e) Fragment Biases: RNA fragmentation creates segments whose starting points were assumed to be located uniformly at random within a transcript. However, it has been demonstrated that there are both positional [42] and sequence-specific [43, 44] biases derived from fragmentation and reverse transcription. Positional bias describes the fact that reads' starting positions are non-uniformly distributed since they are preferentially located toward the transcripts' boundaries. Sequence-specific bias refers to the phenomenon by which the sequences at the reads' boundaries, such as the random hexamer primers used for reverse transcription priming, introduce biases in nucleotide composition and influence the likelihood of being sequenced. Furthermore, fragment length also generates bias since long transcripts result in more reads mapping to them than smaller transcripts. Thereby, for genes with equal levels of expression, the long genes will be overrepresented, distorting the relative expression among genes [45]. Since RNA-Seq read counts are proportional to transcript abundances, expression estimates should be made after fragment bias correction. An effective approach for fragment bias correction has been implemented in the Cufflinks [46] transcriptome assembly and differential expression RNA-Seq analysis tool. The fragment bias correction was based on an algorithm which learns the read sequences and models them as a likelihood function involving abundance and bias parameters such as the probability of finding a fragment with a specific length in a given position [47]. In this manner, bias and expression estimation are performed simultaneously.

1.2.5.3 Read Alignment

In order to determine transcript abundance from reads, it is necessary to align or map reads to a previously assembled reference genome or transcriptome to determine the read's origin. Mapping to a reference genome is more common since it increases the potential information which may be obtained (e.g., identification of novel transcripts and genes) and because many transcriptomes are incomplete. Mapping is a challenging task since RNA-Seq reads are relatively short and they may match non-contiguous regions of the genome due to splice junctions. Furthermore, alignment tools must cope with mismatches and indels caused by genomic variation and sequencing errors. Many alignment tools have been developed. A comprehensive list of alignment tools and their properties was initially published by Fonseca et al. [48] and is kept updated on the Web [49]. A list of some common aligners and their main properties is included in Table 1.1.

The main consideration to be addressed when selecting an aligner is whether RNA-Seq reads span splice junctions. As depicted in Fig. 1.1, unspliced or contiguous alignment tools such as BWA [50], Bowtie [51], and Bowtie2 [52] are useful when mapping reads to a transcriptome, when sequencing microRNAs, or when the organism under study has no intronic regions. Unspliced aligners are thus limited to identifying known exons and do not allow for new splicing event identification. Spliced alignment tools are used when mapping to reference genomes without relying on previously known splice sites. Some of the most commonly used tools for spliced alignment are TopHat [53], TopHat2 [54], Palmapper [55], and STAR [56].

Alignment to a reference genome starts with indexing, the process with which auxiliary structures called indices are created for either the reference sequence or the sequenced reads to allow for faster queries. Indexing the reference genome is more time efficient and thus is used by most alignment tools. Alignment algorithms used for sequencing data analysis are mainly classified into hash tables and suffix trees according to the property of the index used. Hash table indexing was first introduced as an alignment tool by BLAST [67], using a seed and extend approach. In hash table indexing, reads are divided into short *k mer* subsequences called "seeds" and stored in a hash table. The algorithm assumes that at least one "seed" in a read will match the reference. Once a "seed" is aligned, it is extended using more sensitive algorithms such as Smith–Waterman [68] or Needleman–Wunsch [69]. Modifications to hash table indexing algorithms have been performed, and they have been implemented in Novoalign [59], MAQ [65], SHRiMP2 [70], and BFAST [57], among other alignment tools. Suffix trees, on the other hand, are based on the premise that an inexact matching problem may be converted into an exact matching task by constructing a tree (an ordered tree data structure) with all the possible substrings that make up a sequence. The suffix tree data structure enables fast substring searches regardless of sequence size [71]. Among different suffix tree algorithms, one of the most efficient is the FM-index [72] which is based on the Burrows–Wheeler transform (BWT) [73]. BWT is a reversible permutation of characters in a string, and FM-indexing addresses permutations (nodes in a tree) constantly using a backward search. FM-index and BWT, both originally designed for data compression, have been successfully implemented for storing reference genomes and performing rapid queries.

Table 1.1 Overview of common alignment tools

ALIGNERS	Operating system	Language	Alignment algorithm	Input	Output	Paired-end mapping	Splice junction	Read length range	Ref.
BOWTIE	Unix-based, windows	C++	FM-index based on BWT	FAST(A/Q)	SAM, TSV	Yes	No	4 bp–1 k	[51]
BOWTIE2	Unix-based, windows	C++	FM-index based on BWT, dynamic programming	FAST(A/Q)	SAM, TSV	Yes	No	4 bp–5000 k	[52]
PALMapper	Unix-based, web interface	C++	Reference indexing	FAST(A/Q)	SAM, BED (x), SHORE	Yes	Yes	12 bp–12 k	[55]
STAR	Unix-based	C++	Reference indexing	FAST(A/Q)	SAM	Yes	Yes	15 bp–10 k	[56]
BFAST	Unix-based	C	Reference indexing	FAST(A/Q)	SAM, TSV	Yes	No	25–100 bp	[57]
GENOME-MAPPER	Unix-based	C	Reference indexing	FAST (A/Q), SHORE	BED, SHORE	No	No	12 bp–2 k	[58]
NOVAALIGN	Unix-based	C++	Reference indexing	FAST (A/Q), CSFASTA	SAM	Yes	Yes	1–250 bp	[59]
SHRiMP2	Unix-based	Python	Reference indexing	FAST(A/Q)	SAM	Yes	No	30 bp–1 k	[60]
SOAP2	Unix-based	C++	BWT + reference indexing	FAST(A/Q)	SAM/BAM	Yes	No	27 bp–1 k	[61]
MrFAST	Unix-based	C	Reference indexing	FAST(A/Q)	SAM, DIVET	Yes	No	25 bp–1 k	[62]
GNUMAP	Unix-based	C	Reference indexing	FAST(A/Q)	SAM, TSV	No	No	16 bp–1 k	[63]
RMAP	Unix-based	C++	Read indexing	FAST(A/Q)	BED	Yes	No	11 bp–10 k	[64]

(continued)

Table 1.1 (continued)

ALIGNERS	Operating system	Language	Alignment algorithm	Input	Output	Paired-end mapping	Splice junction	Read length range	Ref.
MAQ	Unix-based	C, C++, Perl	Hashing reads	FAST(A/Q)	TSV	Yes	No	8–63 bp	[65]
Mosaik	Unix-based, windows	C++	Reference indexing	FAST(A/Q)	BAM	Yes	No	15 bp–1 k	[66]
BWA	Unix-based, windows	C, C++	FM-index based on BWT	FAST(A/Q)	SAM	Yes	No	4–200 bp	[50]

Reference genome indexes may be built or downloaded as GTF/GFF annotation files most commonly from Ensembl [74] and Illumina iGenomes Web sites [75]. GTF files must be selected carefully using a standard assembly so that chromosome names, gene identifiers, transcription starting sites, and all genomic annotations match between experiments.

Bowtie and Bowtie2 are two of the most efficient unspliced alignment tools because of their low memory requirements and high speed; they both implement an FM-indexing algorithm for achieving ultra-fast alignments. However, neither of these tools are suitable for performing spliced alignments since they cannot align reads when there are large gaps (introns). TopHat addresses spliced alignment limitations by performing a multistep alignment process and using Bowtie as an alignment engine. In the first step, reads are mapped to a reference genome, setting aside reads which were not aligned. In the next step, reads that could not be mapped are broken down into segments and remapped. Finally, reads whose segments were mapped into the same user-defined intronic region are assembled and mapped to that genomic region in an attempt to find splice sites. With this approach, TopHat identifies splice sites without previous splice site annotations and finds novel splicing events [53].

RNA-Seq alignment results are output as SAM/BAM files, and they generally need some further processing such as conversion, sorting, indexing, or merging. SAMTools, implemented in C and Java, is a library for parsing and manipulating alignments prior to downstream analysis [76]. Visualizing aligned reads in a genomic context is recommended for assessing exon coverage, spotting indels and SNPs, displaying splice junctions, identifying novel transcripts, etc. Some available tools for visualization of alignment files are Integrative Genomics Viewer (IGV) [77], Savant [78], and Integrated Genome Browser (IGB) [79].

1.2.5.4 Transcriptome Assembly

In order to quantify gene expression levels from aligned reads, it is necessary to identify which gene isoform generated each read. Therefore, the main aim of transcript assembly is to reconstruct complete transcripts from small overlapping fragments. There are several methods for transcriptome reconstruction, and they can be categorized into genome-guided and genome-independent methods. In genome-guided methods, reads are first mapped to a reference genome and a splicing or exon graph is then constructed for each gene to identify all possible isoforms according to exon combinations. In the splicing graph, each node represents an exon and each connection is an exon junction. Paths that are not evidenced by RNA-Seq reads are eliminated. There are different graph topologies which are implemented to best describe exon combinations for building transcript isoforms. One of the most commonly used tools for genome-guided transcript assembly is Cufflinks, which connects aligned reads based on the location of their spliced alignments [46].

Genome-independent transcriptome reconstruction aims at finding as many long contiguous segments as possible from an assembly graph. The most common strategy is to build a de Bruijn graph, which models overlapped sequence data as a set of *k-mers* (*k* consecutive nucleotides) and their connections [80–82]. Sequences are represented as paths, and branches not supported by reads are eliminated; remaining paths are considered transcript variants. The length of the *k-mer* has an effect on the complexity of the graph, and, although it is conceptually simple, de Bruijn reconstruction approaches have complications such as finding the balance between sensitivity and graph complexity [83]. The value of *k* must be smaller than the read length. However, if *k* is too small, the graph will have excessive connections and will be very sensitive to sequencing errors. If *k* is too large, there must be enough data to make the graph connected. To resolve such issues, several assemblies should be performed with variable values of *k*. Some common de novo assemblers based on de Bruijn graphs are ABySS [84], Trinity [85], Velvet [82], and Oases [86].

1.2.5.5 Expression Quantification

Expression quantification may be performed with respect to transcripts or to genes. Gene expression, the sum of the expression of all its isoforms, is computed by counting reads per gene according to the reference genome's annotation used for mapping. Read counts need to be normalized due to variability introduced by read length bias [45, 87] and due to fluctuations in the number of reads per run [88]. Quantification tools generally output read counts in raw counts, reads per kilobase of transcript per million mapped reads (RPKM), or fragments per kilobase of transcript per million mapped reads (FPKM). RPKM measure normalizes read counts according to the length and to the number of mapped reads per sample [88]. FPKM is used for normalization of paired-end reads since it incorporates dependency estimation [46]. In statistical dependence between two variables (paired-end reads), the levels of one of the variables vary in an exactly determined way with respect to levels of the other variable. All quantification tools are taken as input read alignments in SAM/BAM formats and their reference genome annotation files in GTF/GFF or BED format. They differ in how they handle multimapping reads, which affects expression quantification accuracy [46, 89, 90]. To deal with mapping uncertainty, tools such as Cufflinks use a maximum likelihood function which works by dividing multimapping reads probabilistically according to the abundance of genes they were mapped to [91].

1.2.5.6 Differential Expression Analysis

Often, it is necessary to compare the expression levels of genes or other genomic features between different samples or biological conditions; this is referred to as differential expression analysis. Comparisons are typically performed in a univariate way since it is not possible to fit a multivariate statistical model due to the number of samples being much less than the number of genes.

The ability of detecting differential expression in RNA-Seq experiments depends on the sequencing depth, gene expression, and even on the gene's length, as previously mentioned. A difference in gene expression between two groups is significant only if it is greater than the variability within the group. For estimating variability, biological replicates should be considered. The number of replicates to be used depends on the experiment and the statistical power desired. The purpose of replication is to estimate the variability between and within groups, which is important for hypothesis testing. The set of standards, guidelines, and best practices for RNA-Seq published by the ENCODE Consortium [92] states that two or more biological replicates are sufficient as long as the Pearson correlation of gene expression between them lies between 0.92 and 0.98.

Since RNA-Seq experiments are based on read counts, the initial methods for differential analysis modeled reads as Poisson distributions [46, 87]. However, due to biological variability and the limited number of samples, methods that model count variability as a nonlinear function of mean counts with parametric approaches (e.g., normal, negative binomial distributions) have become popular. Commonly used tools such as DESeq [93], edgeR [94], and Cuffdiff [46] use a negative binomial distribution for modeling RNA-Seq counts. Recently, it has been suggested that RNA-Seq count data may be transformed to apply normal-based microarray-like statistical methods as in the case of Limma software [95]. RNA-Seq data must be normalized transforming counts to have similar empirical distributions across all samples in order to enable comparisons between samples and genes. This step is executed internally by differential expression analysis tools. Table 1.2 makes a comparison of some commonly used differential expression analysis tools. A differential expression analysis should produce a ranked list of differentially expressed genes to be used in downstream applications.

1.2.5.7 Downstream Analysis

Interpretation, visualization, and summarization of differential expression results are important for downstream interpretations. Heat maps and PCA plots are common for finding clusters of differentially expressed genes.

It is of interest to correlate differentially expressed genes to gene sets representing functions, categories, pathways, and others incorporating existing biological knowledge into the analysis. An overrepresentation analysis requires a list of differentially expressed genes which are tested statistically for enrichment in gene sets such as gene ontology (GO) categories, Kyoto Encyclopedia of Genes and Genomes (KEGG), Reactome pathways, and many other databases [96, 97].

Table 1.2 Overview of common tools for differential expression analysis

Properties	Cuffdiff	DESeq	edgeR	baySeq	Limma
Language	C++	R	R	R	R
Operating system	Unix-based OS	Unix-based OS, Windows	Unix-based OS, Windows	Unix-based OS, Windows	Unix-based OS, Windows
Data normalization	Geometric mean, quartile, FPKM	Scaling	Model-based global scaling (TMM)	Scaling, quantile, TMM	Quantile normalization, loess regression, TMM
Read count distribution	Beta negative binomial distribution	Negative binomial distribution	Negative binomial distribution	Negative binomial distribution	Voom transformation of counts into a log distribution
Differential expression test	t-test	Fisher's exact test	Fisher's exact test	Empirical Bayes method to obtain posterior probabilities	Empirical Bayes method
False discovery rate (FDR) estimation method	Benjamini–Hochberg	Benjamini–Hochberg	Benjamini–Hochberg	Bayesian	Benjamini–Hochberg
Reference	[30]	[93]	[94]	[98]	[95]

TMM Trimmed mean of M values

1.3 Characterizing Transcription Factor Regulation by ChIP-chip and ChIP-Seq Methods

The mapping of binding sites for transcription factors (TF), the core transcriptional machinery, and other DNA-binding proteins is essential for understanding gene regulation. Regulatory networks formed by transcription factors and the coordinated activation of their specific target genes play a major role in controlling many cellular processes. The traditional way of constructing gene regulatory networks, via sequential analysis of one or a few genes, is time consuming and labor intensive. Recently, the development of ChIP-chip or ChIP-Seq technology has made it possible to comprehensively identify most in vivo target genes of a given TF at a genome-wide scale, allowing rapid unraveling of signaling pathways [99–101].

In ChIP experiments, TFs are first cross-linked to DNA by treatment with formaldehyde, and chromatin is fragmented to ~300–500 bp fragments. TF-bound chromatins are then immunoprecipitated with specific antibody. Next, the cross-link is reversed by heating to release the precipitated DNA. Immunoprecipitated DNA fragments are hybridized to a microarray (ChIP-chip) or sequenced to generate millions of short sequence tags (ChIP-Seq). Various arrays have been used for ChIP-chip analysis, for example, proximal promoter arrays where about ~ 1 kb PCR products encompassing transcription start sites are used as probes; arrays composed of CpG islands amplified by PCR; large promoter arrays which consist of tiling oligonucleotides of promoter sequences extending up to several kb upstream of the transcription start site; and genome tiling arrays in which a non-repetitive sequence from entire chromosomes is reconstituted using oligonucleotides. As chromosomal sequence is densely covered, higher resolution can be achieved with genome tiling microarrays.

As described previously, sequencing offers various advantages over microarray methods; thus, it has become the predominant technique for profiling genome-wide protein–DNA interactions, chromosomal proteins, and histone marks in vivo [102–104]. For example, the ChIP-Seq assays have higher resolution, lower noise, and better genomic coverage when compared to ChIP-chip assays. Therefore, ChIP-Seq provides more precise mapping of protein-binding sites and sequence motif identification [103, 105]. Several factors influencing ChIP-Seq fidelity need to be addressed.

1.3.1 Analysis of ChIP-Seq Data

The typical output of a ChIP-Seq experiment is a list of millions of short sequence reads. Processing such reads requires filtering and cleaning, mapping to a reference genome, and identification of peak regions. Once significant peaks have been identified, they must be examined, annotated, and associated to a genomic region. The final result is the identification of a transcription factor's motif and binding sites. A general ChIP-Seq workflow is shown in Fig. 1.2. The main issues to consider when analyzing ChIP-Seq data are the following:

(a) Control Sample: ChIP-Seq experiments are prone to artifacts due to effects of DNA shearing and repetitive DNA sequences. DNA shearing during sonication is not uniform because open chromatin regions are fragmented more easily, thus resulting in an uneven distribution of reads. Repetitive DNA sequences may seem enriched when the number of repeats is not considered in calculations. Therefore, the use of a control sample is recommended for peak comparison and significance assessment. Three commonly used control samples are as follows: DNA prior to immunoprecipitation, immunoprecipitated DNA without an antibody, and immunoprecipitated DNA using a non-DNA-binding antibody (e.g., anti-IgG antibody). There is no consensus on which is the most appropriate control.
(b) Sequencing Depth: For a ChIP-Seq analysis to be effective, sufficient genomic coverage, referred to as sequencing depth, is important. However, sequencing

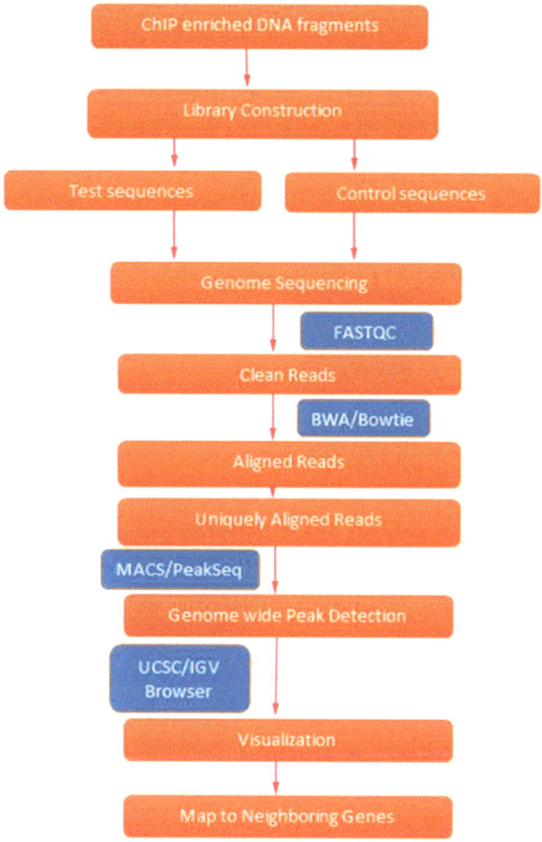

Fig. 1.2 ChIP-Seq workflow. ChIP-Seq experiments allow in vivo determination of where proteins, such as transcription factors, bind to the genome. Bound proteins are cross-linked to chromatin, then fragmented, and immunoprecipitated. ChIP-enriched DNA fragments are used for library construction and sequencing. Reads are filtered according to base quality. Test and control sequences are used for computational mapping to identify genomic locations of bound DNA transcription factors, unveiling potential protein–DNA interactions. Mapped reads are converted into an integer count of "tags." As illustrated, different tools may be used for finding statistically relevant peaks. Finally, the peaks can be visualized and mapped to nearby genes

depth is a potential source of bias since at high sequencing depths, open chromatin regions generate redundant reads which represent false positives [106]. The choice of sequencing depth depends on the genome's size and on the expected number and size of the transcription factor-binding sites. Transcription factors generate highly localized ChIP-Seq signals, and for mammalian genomes, there are thousands of binding sites. For mammalian transcription factors, at least 20 million uniquely mapped reads are currently used for most experiments [107]. For histone marks or proteins with more binding sites such as RNA polymerase II, a higher sequencing depth (e.g., 60 million reads) is needed. To verify whether the sequencing depth was appropriate, a saturation analysis is recommended. Saturation analysis consists of increasing the number of randomly selected reads during read mapping and peak calling for verifying the consistency of peaks called. Saturation analyses are included in some peak caller tools such as SPP [108].

(c) Quality Control Filtering and Read Mapping: Like RNA-Seq data, ChIP-Seq reads must be preprocessed before mapping in order to identify possible sequencing errors and biases. The first filtering criterion is the base calling confidence, computed with the *Phred* quality score for each sequence tag. Low-quality reads should be filtered out and low base quality read ends trimmed. Tools for filtering, trimming, and mapping ChIP-Seq data are the same as for RNA-Seq. After filtering and trimming, the reads are aligned. Alignment/mapping of ChIP-Seq reads is less complex than RNA-Seq reads since large gaps corresponding to splice junctions will not be present. ChIP-Seq aligners generally consider mismatches due to sequencing errors, single nucleotide polymorphisms, and indels. Commonly used mappers are Bowtie [51], BWA [50], SOAP [109], and MAQ [65]. The percentage of uniquely mapped reads must be calculated, and values above 70 % are generally considered normal [110]. However, these values are organism, platform, or protocol dependent. Multimapped reads are most likely due to regions of repeated DNA, and most peak-calling algorithms will filter them out.
Library complexity is the fraction of mapped DNA fragments which are non-redundant, and it may be addressed using the PCR bottleneck coefficient (PBC) from the ENCODE project [111]. PBC computes the fraction of genomic locations with only one unique read mapped against the ones with at least one mapped read. Low-complexity libraries might be due to not enough recovered DNA, resulting in the same PCR-amplified products being sequenced repeatedly. Generally, library complexity is related to antibody quality, over cross-linking, sonication, or over PCR amplification.

(d) Background Signal: Another metric to be considered after mapping is the signal to noise ratio (SNR) of the experiment. During the ChIP-Seq experiment, most of the unbound DNA fragments are washed in the immunoprecipitation step and the library is built with protein–DNA-bound fragments. However, due to nonspecific binding of molecules, non-useful fragments may remain in the library and be sequenced. Such reads become generally spread in the genome and are referred to as background noise and may result in false positives. Noise may be computed from the control sample or modeled with a Poisson or negative binomial distribution.

(e) Peak Calling: One of the most important aims of ChIP-Seq experiments is finding enriched regions in the genome in which more transcription factors (ChIP-Seq tags) were bound to DNA through the number of mapped reads, referred to as "peaks." Several peak callers have been developed, and they mostly differ from each other in the algorithms for signal smoothing and background modeling. Models implemented for statistical assessment of peaks range from Poisson (CSAR [112]), local Poisson (MACS [113]), negative binomial (CisGenome [114]), and even some machine learning techniques, such as hidden Markov models (HPeak [115]). Peak callers report a p value or false discovery rate (FDR) as an enrichment metric which is greatly affected by variables, such as sequencing depth, real number of binding sites, and the statistical model used. There is no consensus on how to best estimate the best FDR value for ChIP-Seq experiments. Table 1.3 lists the main characteristics of some commonly used peak callers.

(f) Reproducibility: It is recommended to develop experiments with at least two biological replicates for verifying reproducibility of reads and identified peaks [107]. The reproducibility of reads can be computed with metrics such as Pearson correlation coefficient of mapped read counts at each genomic site [116].

(g) Downstream Analysis: Once significant and reproducible peaks have been found, it is necessary to associate them with relevant genomic regions, such as transcription start sites, gene promoters, and intergenic regions. It is common to view the identified peaks and reads in a genome browser to examine regions of interest. Generally, peaks are uploaded as BED or GFF file formats and reads with WIG file format. HOMER or BEDTools may be used to calculate distances from peaks to landmark regions (e.g., genes). The most common downstream analysis of a ChIP-seq experiment is the discovery of binding sequence motifs [117]. The read sequences of the top-scoring peaks can be entered in FASTA format into motif discovery programs such as MEME [118] resulting in motif discovery, enrichment, and location analysis.

The in vivo binding targets of TFs identified above can be further correlated with the differentially expressed genes using Gene Set Enrichment Analysis (GSEA) software [101, 123]. The factors that show enriched binding to the differentially expressed genes can be selected for further genetic testing. Finally, to understand the intricate relationship of the TFs that are differentially expressed, one can construct a network among coregulated TFs and incorporate ChIP-Seq result into the network. Thus, the underlying regulatory mechanism can be revealed, such as autoregulation (where a factor interacts with its own promoter region), cross-factor control (where pairs of factors directly bind each other's promoter regions), and positive/negative feedback loop.

Table 1.3 Overview of common peak callers used for ChIP-Seq data analysis

Software	Background model	Signal profile creation method	False discovery rate (FDR) estimation method	Statistical model for peak identification	Ref.
SPP	Poisson	Window scan	Ratio of significant scores between sample and control	Poisson model	[108]
CisGenome	Negative binomial/control sample	Window scan	Ratio between expected to observed peak number	Conditional/negative binomial model	[114]
PeakSeq	Local Poisson/control sample	Extended tag aggregation	Benjamini–Hochberg	Conditional binomial model	[119]
MACS	Local Poisson/control sample	Shift tags + window scan	Ratio between number of peaks in control and in ChIP	Local Poisson model	[113]
QuEST	Control sample	Kernel density estimation	Based on control sample	Threshold-based model	[120]
FindPeaks	Uniform	Overlap-based	Monte Carlo simulation	Peak height threshold	[121]
F-Seq	Kernel density estimation	Kernel density estimation	None	Peak height threshold	[122]

1.4 Integrated Study of Gene Expression and Transcriptional Regulation—An Example: Identification of Key Factors Regulating Self-renewal and Differentiation in EML Hematopoietic Precursor Cells by RNA-Seq and ChIP-Seq Analyses

1.4.1 The Multipotential EML Cell Line Is a Favorable System to Study the Control of Early Hematopoietic Self-renewal and Differentiation

The hematopoietic system has provided a leading model for stem cell studies, and there is great interest in elucidating the mechanisms that control the decision of HSC self-renewal and differentiation [124–130]. This switch is important for understanding hematopoietic diseases and manipulating HSCs for therapeutic purposes. However, because HSCs are currently unable to proliferate extensively in vitro, this severely limits the types of biochemical analyses that can be performed, and consequently, the mechanisms that control the decision between early-stage HSC self-renewal and differentiation remain unclear [131].

The mouse (*Mus musculus*) EML (erythroid, myeloid, and lymphocytic) multipotential hematopoietic precursor cell is an ideal system for studying the molecular control of early hematopoietic differentiation events. EML cells are derived from mouse bone marrow cells and are cultured in the presence of stem cell factor (SCF). These cells can be rederived or repeatedly cloned, and still retain their multipotentiality [132–134]. The ability of EML cells to propagate extensively in medium containing SCF makes them ideal for biochemical and genetic assays, as well as for high-throughput functional screens [126, 135]. Phenotypically, EML cells express many of the cell surface markers' characteristic of hematopoietic progenitor cells, including SCA1, CD34, and c-KIT. Functionally, when treated with different growth factors, such as SCF, IL-3, GM-CSF, and EPO, EML cells can differentiate into distinct cell lineages including B-lymphocyte, erythrocyte, neutrophil, macrophage, mast cell, and megakaryocyte lineages [132].

Interestingly, in culture, the Lin-SCA+ CD34+ subpopulation of EML cells gives rise to a mixed population containing similar numbers of self-renewing Lin-SCA+ CD34+ precursor cells and partially differentiated Lin-SCA-CD34− cells (henceforth referred to as CD34+ and CD34− cells, respectively) [136]. Although the two populations resemble each other morphologically, only the CD34+ population propagates in SCF-containing media, while the growth of CD34− cells requires the cytokine IL-3 [136]. The closest normal analogs of CD34+ cells are short-term (ST) HSC or multipotent progenitors (MPP). Similar to short-term (ST) HSC, CD34+ cells are capable of self-renewal; like MPP, when treated with cytokines such as IL-3, CD34+ cells can give rise to CD34− cells with more restricted potential. A number of erythroid genes, such as α- and β-hemoglobin, Gata1, Epor (erythropoietin receptor), and Eraf (erythroid associated factor), as well as mast cell proteases are expressed at a significantly higher level in the CD34− cell population than in CD34+ cells [136, 137]. This indicates that the CD34− cells were, at minimum, differentiated into a state with prominent erythroid potential.

The ability of CD34+ cells to both differentiate and self-renew in suspension culture, in the absence of any anatomical niche or other cell types, suggests that CD34+ cells are regulated by a tightly controlled endogenous mechanism that guides the generation of the variety and relative abundance of the cell types in culture. Understanding the molecular events that regulate the transition between the two types of putative precursors in the EML multipotent hematopoietic cell line will give insights into the fundamental mechanisms of autonomous and balanced cell fate choice available to stem cells and intermediate-stage cancer precursor cells [126].

1.4.2 Mapping Transcription Regulatory Networks and Identifying Master Regulators

The regulatory inputs and functional outputs of various downstream genes constitute network-like architectures [138]. The linkage relationships in a complex network provide causal clues about how a specific eukaryotic process is regulated at the molecular level. Using these methods, regulatory networks have been constructed for the yeast cell cycle [139–141], yeast development [141, 142], and human embryonic stem cell self-renewal [99]. For example, in the study of yeast pseudohyphal development, the binding targets of six key transcription factors (Ste12, Tec1, Sok2, Phd1, Mga1, and Flo8) were identified. The binding network formed by these factors and their target genes were analyzed, and Mga1 and Phd1 were found to be the targets of many factors in the network. These factors were called target hubs [142].

It appears that target hubs are especially likely to be "master regulators." Master regulators have been identified as transcription factors whose ectopic expression alone is capable of activating a biological pathway. For example, MyoD is capable of activating a terminal muscle differentiation program in primary cells and in differentiated cell lines [143]. The target hubs Mga1 and Phd1 also appear to be such "master regulators," serving as key nodal points that orchestrate a large number of regulatory inputs into a complex response [144–146]. Overexpression of either of these target hub proteins under conditions that do not normally activate the pseudohyphal response specifically induces this process. The distinct nature of the master regulators allows us to use them as a switch to control cellular processes, which has important therapeutic applications.

1.4.3 Identifying Key Factors Regulating Self-renewal and Differentiation

In order to identify the "switch" in cell self-renewal and differentiation, we constructed regulatory circuits controlling early hematopoietic differentiation by using the gene expression and ChIP-Seq data. We examined transcription factors that were significantly upregulated in CD34+ cells relative to CD34− cells using RNA-Seq and found *Tcf7* (also referred to by the symbol *Tcf1*) to be the most strongly upregulated transcription factor (Fig. 1.3) [27].

The binding motifs of the TCF family of transcription factors are significantly enriched among genes that are expressed at a higher level in CD34+ than in CD34−

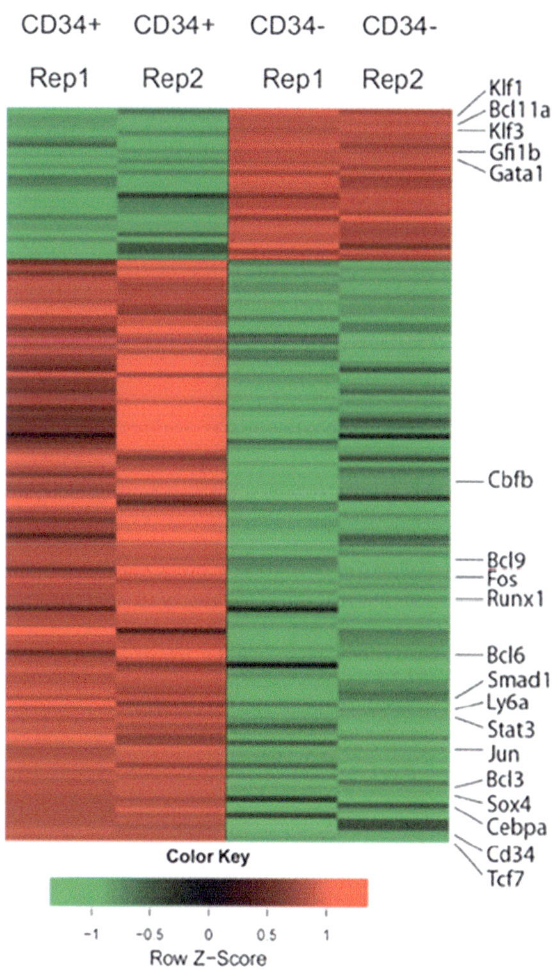

Fig. 1.3 Heat map of differentially expressed transcription factors (>1.5-fold) between Lin-CD34+ cells and Lin-CD34− cells. Two replicates were shown for each cell type. *Red color* represents upregulated genes and *green color* represents downregulated genes. Genes mentioned in the text are labeled. CD34 and Ly6a (Sca1) are cell surface markers. Adapted from Wu et al. [27]

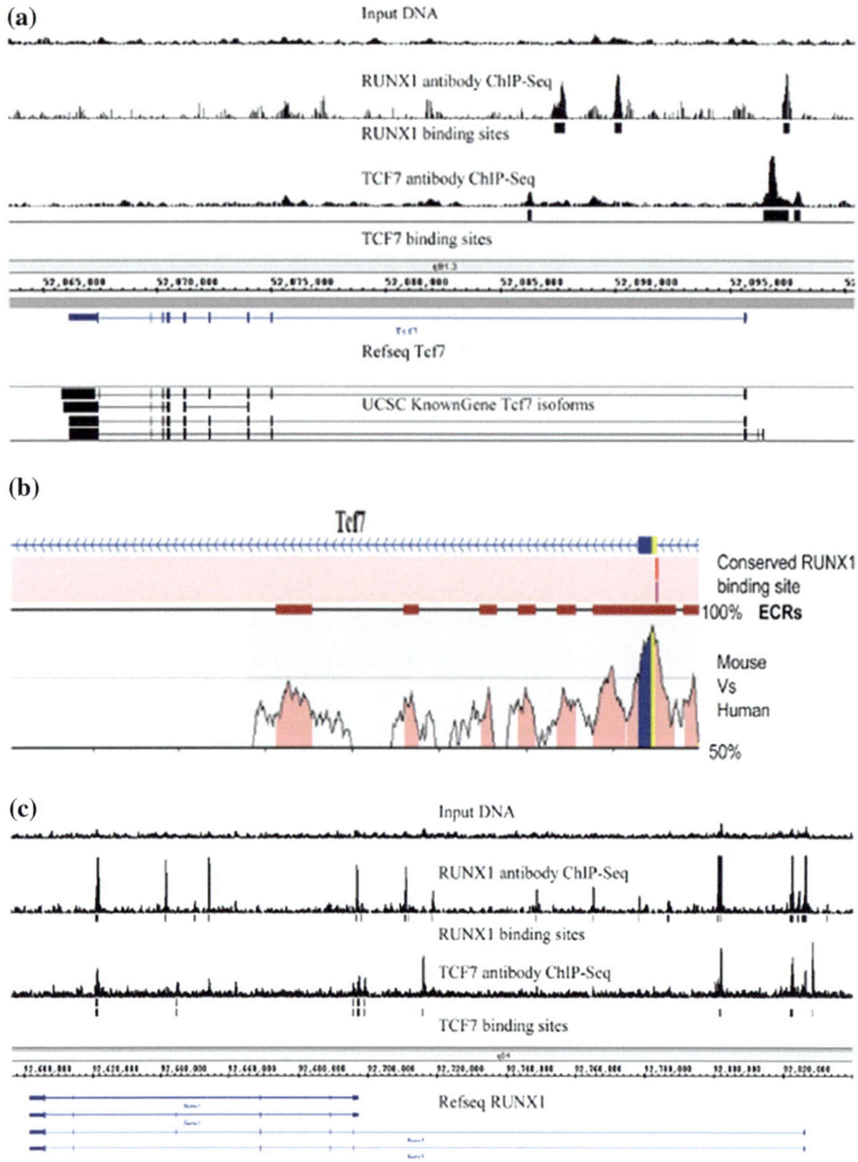
(a)
Input DNA
RUNX1 antibody ChIP-Seq
RUNX1 binding sites
TCF7 antibody ChIP-Seq
TCF7 binding sites
Refseq Tcf7
UCSC KnownGene Tcf7 isoforms
(b)
Tcf7
Conserved RUNX1 binding site
100% ECRs
Mouse Vs Human
50%
(c)
Input DNA
RUNX1 antibody ChIP-Seq
RUNX1 binding sites
TCF7 antibody ChIP-Seq
TCF7 binding sites
Refseq RUNX1

◂ **Fig. 1.4** Identification of transcription factor-binding targets using ChIP sequencing. **a** Tcf7 is bound by both itself and by RUNX1 (AML1). Peaks indicate ChIP sequencing signal. Input genomic DNA serves as the negative control. The "binding sites" tracks (*black vertical bars*) show the transcription factor-binding loci determined using the PeakSeq program (normalized against genomic input DNA; q-value 0.001). Data are visualized in Integrated Genome Browser. **b** Identification of evolutionarily conserved RUNX1-binding sites at Tcf7 promoter region using REGULATORY VISTA. The graph shows conserved and aligned AML1/RUNX1 transcription factor-binding sites between mouse and human genomes using a matrix similarity score of 1 (the mt stringent). Two versions of the AML1-binding sites were found (AML1 and AML_Q6). The *ECRs: Evolutionarily conserved regions are indicated by deep *red blocks*. The degree of conservation (50–100 %) is indicated by the height of the peaks. Coding region is shown in *blue*, and UTR is shown in *yellow*. **c** Runx1 promoter is bound by both TCF7 and itself. Adapted from Wu et al. [27]

cells [27]. Therefore, we hypothesize that there are key regulators in transcriptional regulatory networks that determine the choice between EML cell self-renewal and differentiation, and TCF7 is one of the key transcription factors.

Subsequently, we identified in vivo binding targets of TCF7 using ChIP-Seq [27]. We found that TCF7 binds to its own promoter and the promoter of *Runx1 (Aml1)*, a developmental determinant in hematopoietic cells that is best known for its critical role in haematological malignancies [147, 148] (Fig. 1.4). We showed that TCF7 and RUNX1 (AML1) bind to each other's promoter regions, and a large number of common target genes are bound by RUNX1 and TCF7. TCF7 is necessary for the production of the short isoforms, but not the long isoforms of RUNX1, suggesting that TCF7 and the short isoforms of RUNX1 function coordinately in regulation. TCF7 knockdown experiments and Gene Set Enrichment Analyses suggest that TCF7 plays a dual role in promoting the expression of genes characteristic of self-renewing CD34+ cells while repressing genes activated in the partially differentiated CD34− state. Finally, through network analysis, we found that TCF7 and RUNX1 bind and regulate a network of upregulated transcription factors in the CD34+ cells which characterize the self-renewal property of the CD34+ cells, including Stat3, Sox4, F, Scl/Tal1, Etv6/Tel, Ppard, Smads, Cebpa, Gfi1, and Fli-1 (Fig. 1.5).

In summary, our results elucidated novel components and mechanisms that control the renewal and differentiation of hematopoietic precursor cells. The elucidation of the networks suggested potential master regulators that control early hematopoietic differentiation. Genetic manipulation of the master regulators may reveal how to induce hematopoietic precursor cell self-renewal in vitro or reprogram partially differentiated hematopoietic precursor cells back to a self-renewing state. Increasing the long-term ability of human hematopoietic precursor cells to reconstitute bone marrow is highly relevant for the therapy of leukemia and regenerative medicine.

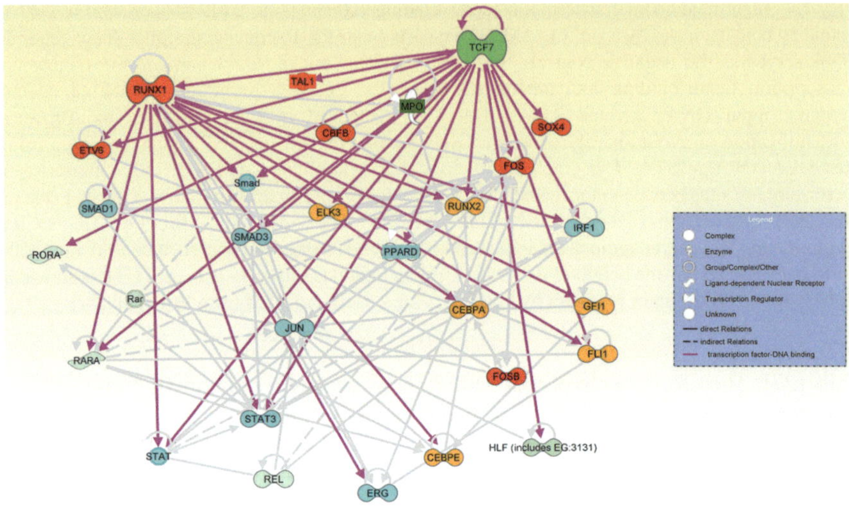

Fig. 1.5 Transcription factors TCF7 together with RUNX1 regulate a transcriptional regulatory network. The network involved in HSC establishment and development (*red nodes*), cell growth control (*blue nodes*), and multipotency (*orange nodes*) was identified among upregulated genes in CD34+ cells (twofold) and displayed by Ingenuity Pathway Analysis software (IPA). *Gray lines* are IPA-annotated relations based on the literature. *Pink lines* indicate TCF7 or RUNX1 binding to gene targets that were identified by our ChIP-Seq experiments. The shades of *green color* of the nodes in the network indicate the level of upregulation in CD34+ cells. Sox4, Mpo, Tal1, and Ppard were TCF7-binding targets that were added to the network manually because of their obvious interesting function in hematopoiesis and self-renewal. All other nodes were from default IPA analysis. Direct relations were indicated by *solid line* or *arrows*. Indirect relations were indicated by *dotted line*. Please see Ingenuity Pathway Analysis software (IPA, https://analysis.ingenuity.com/) Online Help section for detailed definitions. Adapted from Wu et al [27]

1.5 Future Prospective

The advancement of sequencing technology and computational analyses has greatly increased our knowledge of gene transcription and regulation. However, many challenges still remain. Difficulties in deciphering the anatomy of mammalian genes exist at multiple levels, including topics discussed in later chapters, such as complicated RNA, large amounts of intervening (noncoding) sequences, and the imperfection of computational algorithms. Additional issues include overlapping reading frames of protein-coding genes, antisense transcriptional units, the situation where the exon of one gene is encoded within the intron of another, and pseudogenes [149–151]. It will be impossible to find all genes and regulatory elements solely by analyzing genomic nucleotide sequences. Therefore, the eventual solution of annotation lies in large-scale systematic functional genomics experiments and conservation information from cross-genome comparisons [152].

Additionally, we should always take caution when interpreting data from a single kind of "omic" approach. For example, we cannot immediately conclude that a protein is expressed at a higher level from an upregulated signal by using microarray or RNA-Seq alone. Integrating data obtained from multiple distinct approaches will make conclusions more reliable. Theoretically, as multiple "omic" functional maps are overlaid, genes involved in the same process will cocluster in various maps. There are many challenges ahead in developing statistical and computational strategies for integrating these data, for improving annotation, and for making them available to the scientific community. The long-term goal is to understand the intricate and dynamic functional relationships between all components involved in particular biological processes as a whole, in order to be able to predict the potential behaviors of these systems in response to perturbations and thus be able to restore. This approach will provide answers for treating diseases.

Acknowledgments We thank Dr. Eva Zsigmond for reading and editing our manuscript. JQW, RCDD, and SM are supported by grants from the National Institutes of Health R01 NS088353 and R00 HL093213, the Staman Ogilvie Fund—Memorial Hermann Foundation, Mission Connect—a program of the TIRR Foundation, the Senator Lloyd & B.A. Bentsen Center for Stroke Research, UTHealth BRAIN Initiative and CTSA UL1 TR000371, and a grant from the University of Texas System Neuroscience and Neurotechnology Research Institute (Grant #362469).

References

1. Liang F, Holt I, Pertea G, Karamycheva S, Salzberg SL, Quackenbush J. Gene index analysis of the human genome estimates approximately 120,000 genes. Nat Genet. 2000;25(2): 239–40.
2. Adams MD, Kelley JM, Gocayne JD, Dubnick M, Polymeropoulos MH, Xiao H, et al. Complementary DNA sequencing: expressed sequence tags and human genome project. Science. 1991;252(5013):1651–6.
3. Wolfsberg TG, Landsman D. A comparison of expressed sequence tags (ESTs) to human genomic sequences. Nucleic Acids Res. 1997;25(8):1626–32.
4. Bailey LC Jr, Searls DB, Overton GC. Analysis of EST driven gene annotation in human genomic sequence. Genome Res. 1998;8(4):362–76.
5. Das M, Burge CB, Park E, Colinas J, Pelletier J. Assessment of the total number of human transcription units. Genomics. 2001;77(1–2):71–8.
6. Velculescu VE, Zhang L, Vogelstein B, Kinzler KW. Serial analysis of gene expression. Science. 1995;270(5235):484–7.
7. Saha S, Sparks AB, Rago C, Akmaev V, Wang CJ, Vogelstein B, et al. Using the transcriptome to annotate the genome. Nat Biotechnol. 2002;20(5):508–12.
8. Wei CL, Ng P, Chiu KP, Wong CH, Ang CC, Lipovich L, et al. 5′ Long serial analysis of gene expression (LongSAGE) and 3′ LongSAGE for transcriptome characterization and genome annotation. Proc Natl Acad Sci USA. 2004;101(32):11701–6 (Epub 2004/07/24).
9. Kawai J, Shinagawa A, Shibata K, Yoshino M, Itoh M, Ishii Y, et al. Functional annotation of a full-length mouse cDNA collection. Nature. 2001;409(6821):685–90.
10. Clark MD, Hennig S, Herwig R, Clifton SW, Marra MA, Lehrach H, et al. An oligonucleotide fingerprint normalized and expressed sequence tag characterized zebrafish cDNA library. Genome Res. 2001;11(9):1594–602.

11. Okazaki Y, Furuno M, Kasukawa T, Adachi J, Bono H, Kondo S, et al. Analysis of the mouse transcriptome based on functional annotation of 60,770 full-length cDNAs. Nature. 2002;420(6915):563–73.
12. Stapleton M, Carlson J, Brokstein P, Yu C, Champe M, George R, et al. A Drosophila full-length cDNA resource. Genome Biol. 2002;3(12):RESEARCH0080.
13. Reboul J, Vaglio P, Rual JF, Lamesch P, Martinez M, Armstrong CM, et al. *C. elegans* ORFeome version 1.1: experimental verification of the genome annotation and resource for proteome-scale protein expression. Nat Genet. 2003;34(1):35–41.
14. Wiemann S, Weil B, Wellenreuther R, Gassenhuber J, Glassl S, Ansorge W, et al. Toward a catalog of human genes and proteins: sequencing and analysis of 500 novel complete protein coding human cDNAs. Genome Res. 2001;11(3):422–35.
15. Ota T, Suzuki Y, Nishikawa T, Otsuki T, Sugiyama T, Irie R, et al. Complete sequencing and characterization of 21,243 full-length human cDNAs. Nat Genet. 2004;36(1):40–5.
16. Gerhard DS, Wagner L, Feingold EA, Shenmen CM, Grouse LH, Schuler G, et al. The status, quality, and expansion of the NIH full-length cDNA project: the Mammalian Gene Collection (MGC). Genome Res. 2004;14(10B):2121–7.
17. Strausberg RL, Feingold EA, Grouse LH, Derge JG, Klausner RD, Collins FS, et al. Generation and initial analysis of more than 15,000 full-length human and mouse cDNA sequences. Proc Natl Acad Sci USA. 2002;99(26):16899–903.
18. Temple G, Gerhard DS, Rasooly R, Feingold EA, Good PJ, Robinson C, et al. The completion of the Mammalian Gene Collection (MGC). Genome Res. 2009;19(12):2324–33.
19. Bareyre FM, Schwab ME. Inflammation, degeneration and regeneration in the injured spinal cord: insights from DNA microarrays. Trends Neurosci. 2003;26:555–63.
20. Carmel JB, Galante a, Soteropoulos P, Tolias P, Recce M, Young W, et al. Gene expression profiling of acute spinal cord injury reveals spreading inflammatory signals and neuron loss. Physiol Genomics 2001;7:201–13.
21. Velardo MJ, Burger C, Williams PR, Baker HV, López MC, Mareci TH, et al. Patterns of gene expression reveal a temporally orchestrated wound healing response in the injured spinal cord. J Neurosci. 2004;24:8562–76.
22. Liu CL, Jin AM, Tong BH. Detection of gene expression pattern in the early stage after spinal cord injury by gene chip. Chin J Traumatol. 2003;6(1):18–22 (Epub 2003/01/25).
23. Tachibana T, Noguchi K, Ruda MA. Analysis of gene expression following spinal cord injury in rat using complementary DNA microarray. Neurosci Lett. 2002;327(2):133–7 (Epub 2002/07/06).
24. Jaerve A, Kruse F, Malik K, Hartung HP, Muller HW. Age-dependent modulation of cortical transcriptomes in spinal cord injury and repair. PLoS One. 2012;7(12):e49812 (Epub 2012/12/14).
25. Kahvejian A, Quackenbush J, Thompson JF. What would you do if you could sequence everything? Nat Biotechnol. 2008;26:1125–33.
26. Wu JQ, Habegger L, Noisa P, Szekely A, Qiu C, Hutchison S, et al. Dynamic transcriptomes during neural differentiation of human embryonic stem cells revealed by short, long, and paired-end sequencing. Proc Natl Acad Sci USA. 2010;107(11):5254–9 (Epub 2010/03/03).
27. Wu JQ, Seay M, Schulz, V., Hariharan, M., Tuck, D., Lian, J., Du, J., Shi, M., Ye, Z.J, Gerstein M, Snyder M, Weissman S. TCF7 is a key regulator of the self-renewal and differentiation switch in a multipotential hematopoietic cell line. PLoS Genet. 2012;8(3): e1002565 (Epub 2012).
28. Wang Z, Gerstein M, Snyder M. RNA-Seq: a revolutionary tool for transcriptomics. Nat Rev Genet. 2009;10(1):57–63 (Epub 2008/11/19).
29. Torres-Garcia W, Zheng S, Sivachenko A, Vegesna R, Wang Q, Yao R, et al. PRADA: pipeline for RNA sequencing data analysis. Bioinformatics. 2014;30(15):2224–6 (Epub 2014/04/04).
30. Trapnell C, Roberts A, Goff L, Pertea G, Kim D, Kelley DR, et al. Differential gene and transcript expression analysis of RNA-seq experiments with TopHat and Cufflinks. Nat Protoc. 2012;7(3):562–78 (Epub 2012/03/03).

31. Kalari KR, Nair AA, Bhavsar JD, O'Brien DR, Davila JI, Bockol MA, et al. MAP-RSeq: mayo analysis pipeline for RNA sequencing. BMC Bioinf. 2014;15:224 (Epub 2014/06/29).
32. Cumbie JS, Kimbrel JA, Di Y, Schafer DW, Wilhelm LJ, Fox SE, et al. GENE-counter: a computational pipeline for the analysis of RNA-Seq data for gene expression differences. PLoS ONE. 2011;6(10):e25279 (Epub 2011/10/15).
33. Fonseca NA, Marioni J, Brazma A. RNA-Seq gene profiling–a systematic empirical comparison. PLoS ONE. 2014;9(9):e107026 (Epub 2014/10/01).
34. Schmieder R, Edwards R. Quality control and preprocessing of metagenomic datasets. Bioinformatics. 2011;27(6):863–4 (Epub 2011/02/01).
35. Cock PJ, Fields CJ, Goto N, Heuer ML, Rice PM. The Sanger FASTQ file format for sequences with quality scores, and the Solexa/Illumina FASTQ variants. Nucleic Acids Res. 2010;38(6):1767–71 (Epub 2009/12/18).
36. Ewing B, Green P. Base-calling of automated sequencer traces using phred. II. Error probabilities. Genome Res. 1998;8(3):186–94 (Epub 1998/05/16).
37. Del Fabbro C, Scalabrin S, Morgante M, Giorgi FM. An extensive evaluation of read trimming effects on Illumina NGS data analysis. PLoS ONE. 2013;8(12):e85024 (Epub 2014/01/01).
38. Bolger AM, Lohse M, Usadel B. Trimmomatic: a flexible trimmer for Illumina sequence data. Bioinformatics. 2014;30(15):2114–20 (Epub 2014/04/04).
39. Martin M. Cutadapt removes adapter sequences from high-throughput sequencing reads. EMBnet J. 2011;17:10–2.
40. Smeds L, Kunstner A. ConDeTri–a content dependent read trimmer for Illumina data. PLoS ONE. 2011;6(10):e26314 (Epub 2011/11/01).
41. Hansen KD, Irizarry RA, Wu Z. Removing technical variability in RNA-seq data using conditional quantile normalization. Biostatistics. 2012;13(2):204–16 (Epub 2012/01/31).
42. Bohnert R, Ratsch G. rQuant.web: a tool for RNA-Seq-based transcript quantitation. Nucleic Acids Res. 2010;38(Web Server issue):W348-51 (Epub 2010/06/17).
43. Srivastava S, Chen L. A two-parameter generalized Poisson model to improve the analysis of RNA-seq data. Nucleic Acids Res. 2010;38(17):e170 (Epub 2010/07/31).
44. Hansen KD, Brenner SE, Dudoit S. Biases in Illumina transcriptome sequencing caused by random hexamer priming. Nucleic Acids Res. 2010;38(12):e131 (Epub 2010/04/17).
45. Oshlack A, Wakefield MJ. Transcript length bias in RNA-seq data confounds systems biology. Biol Direct. 2009;4:14 (Epub 2009/04/18).
46. Trapnell C, Williams BA, Pertea G, Mortazavi A, Kwan G, van Baren MJ, et al. Transcript assembly and quantification by RNA-Seq reveals unannotated transcripts and isoform switching during cell differentiation. Nat Biotechnol. 2010;28(5):511–5 (Epub 2010/05/04).
47. Roberts A, Trapnell C, Donaghey J, Rinn JL, Pachter L. Improving RNA-Seq expression estimates by correcting for fragment bias. Genome Biol. 2011;12(3):R22 (Epub 2011/03/18).
48. Fonseca NA, Rung J, Brazma A, Marioni JC. Tools for mapping high-throughput sequencing data. Bioinformatics. 2012;28(24):3169–77 (Epub 2012/10/13).
49. Updated listing of mappers. Available from: http://wwwdev.ebi.ac.uk/fg/hts_mappers/.
50. Li H, Durbin R. Fast and accurate short read alignment with Burrows-Wheeler transform. Bioinformatics. 2009;25(14):1754–60 (Epub 2009/05/20).
51. Langmead B, Trapnell C, Pop M, Salzberg SL. Ultrafast and memory-efficient alignment of short DNA sequences to the human genome. Genome Biol. 2009;10(3):R25 (Epub 2009/03/06).
52. Langmead B, Salzberg SL. Fast gapped-read alignment with Bowtie 2. Nat Methods. 2012;9 (4):357–9 (Epub 2012/03/06).
53. Trapnell C, Pachter L, Salzberg SL. TopHat: discovering splice junctions with RNA-Seq. Bioinformatics. 2009;25(9):1105–11 (Epub 2009/03/18).
54. Kim D, Pertea G, Trapnell C, Pimentel H, Kelley R, Salzberg SL. TopHat2: accurate alignment of transcriptomes in the presence of insertions, deletions and gene fusions. Genome Biol. 2013;14(4):R36 (Epub 2013/04/27).

55. Jean G, Kahles A, Sreedharan VT, De Bona F, Ratsch G. RNA-Seq read alignments with PALMapper. Current protocols in bioinformatics/editoral board, Andreas D Baxevanis [et al]. 2010;Chapter 11:Unit 11 6 (Epub 2010/12/15).
56. Dobin A, Davis CA, Schlesinger F, Drenkow J, Zaleski C, Jha S, et al. STAR: ultrafast universal RNA-seq aligner. Bioinformatics. 2013;29(1):15–21 (Epub 2012/10/30).
57. Homer N, Merriman B, Nelson SF. BFAST: an alignment tool for large scale genome resequencing. PLoS ONE. 2009;4(11):e7767 (Epub 2009/11/13).
58. Schneeberger K, Hagmann J, Ossowski S, Warthmann N, Gesing S, Kohlbacher O, et al. Simultaneous alignment of short reads against multiple genomes. Genome Biol. 2009;10(9): R98 (Epub 2009/09/19).
59. Novocraft. 2010. Available from: http://www.novocraft.com/.
60. David M, Dzamba M, Lister D, Ilie L, Brudno M. SHRiMP2: sensitive yet practical SHort read mapping. Bioinformatics. 2011;27(7):1011–2 (Epub 2011/02/01).
61. Li R, Yu C, Li Y, Lam TW, Yiu SM, Kristiansen K, et al. SOAP2: an improved ultrafast tool for short read alignment. Bioinformatics. 2009;25(15):1966–7 (Epub 2009/06/06).
62. Alkan C, Kidd JM, Marques-Bonet T, Aksay G, Antonacci F, Hormozdiari F, et al. Personalized copy number and segmental duplication maps using next-generation sequencing. Nat Genet. 2009;41(10):1061–7 (Epub 2009/09/01).
63. Clement NL, Clement MJ, Snell Q, Johnson WE. Parallel mapping approaches for GNUMAP. IPDPS. 2011;435–43 (Epub 2011/01/01).
64. Smith AD, Chung WY, Hodges E, Kendall J, Hannon G, Hicks J, et al. Updates to the RMAP short-read mapping software. Bioinformatics. 2009;25(21):2841–2 (Epub 2009/09/09).
65. Li H, Ruan J, Durbin R. Mapping short DNA sequencing reads and calling variants using mapping quality scores. Genome Res. 2008;18(11):1851–8 (Epub 2008/08/21).
66. Maher MC, Hernandez RD. Rock, paper, scissors: harnessing complementarity in ortholog detection methods improves comparative genomic inference. G3 (Bethesda). 2015;5(4): 629–38 (Epub 2015/02/26).
67. Altschul SF, Gish W, Miller W, Myers EW, Lipman DJ. Basic local alignment search tool. J Mol Biol. 1990;215(3):403–10 (Epub 1990/10/05).
68. Smith TF, Waterman MS. Identification of common molecular subsequences. J Mol Biol. 1981;147(1):195–7 (Epub 1981/03/25).
69. Needleman SB, Wunsch CD. A general method applicable to the search for similarities in the amino acid sequence of two proteins. J Mol Biol. 1970;48(3):443–53 (Epub 1970/03/01).
70. Rumble SM, Lacroute P, Dalca AV, Fiume M, Sidow A, Brudno M. SHRiMP: accurate mapping of short color-space reads. PLoS Comput Biol. 2009;5(5):e1000386 (Epub 2009/05/23).
71. Barsky M, Stege U, Thomo A, Upton C, editors. Suffix trees for very large genomic sequences. CIKM '09: Proceedings of the 18th ACM Conference on Information and Knowledge Management; 2009; New York, NY, USA.
72. Ferragina P, Manzini G, editors. Opportunistic data structures with applications. Proceedings of the 41st Symposium on Foundations of Computer Science (FOCS 2000); 2000; Redondo Beach, CA.
73. Burrows M, Wheeler D. A block sorting lossless data compression algorithm. Palo Alto, CA: Digital Equipment Corporation; 1994.
74. Cunningham F, Amode MR, Barrell D, Beal K, Billis K, Brent S, et al. Ensembl 2015. Nucleic Acids Res. 2015;43(Database issue):D662-9 (Epub 2014/10/30).
75. iGenomes. Available from: https://support.illumina.com/sequencing/sequencing_software/igenome.html.
76. Li H, Handsaker B, Wysoker A, Fennell T, Ruan J, Homer N, et al. The sequence alignment/map format and SAMtools. Bioinformatics. 2009;25(16):2078–9 (Epub 2009/06/10).

77. Thorvaldsdottir H, Robinson JT, Mesirov JP. Integrative genomics viewer (IGV): high-performance genomics data visualization and exploration. Brief Bioinform. 2013;14(2):178–92 (Epub 2012/04/21).
78. Fiume M, Smith EJ, Brook A, Strbenac D, Turner B, Mezlini AM, et al. Savant Genome Browser 2: visualization and analysis for population-scale genomics. Nucleic Acids Res. 2012;40(Web Server issue):W615-21 (Epub 2012/05/29).
79. Nicol JW, Helt GA, Blanchard SG Jr, Raja A, Loraine AE. The integrated genome browser: free software for distribution and exploration of genome-scale datasets. Bioinformatics. 2009;25(20):2730–1 (Epub 2009/08/06).
80. Surget-Groba Y, Montoya-Burgos JI. Optimization of de novo transcriptome assembly from next-generation sequencing data. Genome Res. 2010;20(10):1432–40 (Epub 2010/08/10).
81. De Bruijn NG. A combinatorial problem. Koninklijke Nederlandse Akademie v Wetenschappen. 1946;46(6).
82. Zerbino DR, Birney E. Velvet: algorithms for de novo short read assembly using de Bruijn graphs. Genome Res. 2008;18(5):821–9 (Epub 2008/03/20).
83. Robertson G, Schein J, Chiu R, Corbett R, Field M, Jackman SD, et al. De novo assembly and analysis of RNA-seq data. Nat Methods. 2010;7(11):909–12 (Epub 2010/10/12).
84. Simpson JT, Wong K, Jackman SD, Schein JE, Jones SJ, Birol I. ABySS: a parallel assembler for short read sequence data. Genome Res. 2009;19(6):1117–23 (Epub 2009/03/03).
85. Grabherr MG, Haas BJ, Yassour M, Levin JZ, Thompson DA, Amit I, et al. Full-length transcriptome assembly from RNA-Seq data without a reference genome. Nat Biotechnol. 2011;29(7):644–52 (Epub 2011/05/17).
86. Schulz MH, Zerbino DR, Vingron M, Birney E. Oases: robust de novo RNA-seq assembly across the dynamic range of expression levels. Bioinformatics. 2012;28(8):1086–92 (Epub 2012/03/01).
87. Marioni JC, Mason CE, Mane SM, Stephens M, Gilad Y. RNA-seq: an assessment of technical reproducibility and comparison with gene expression arrays. Genome Res. 2008;18(9):1509–17 (Epub 2008/06/14).
88. Mortazavi A, Williams BA, McCue K, Schaeffer L, Wold B. Mapping and quantifying mammalian transcriptomes by RNA-Seq. Nat Methods. 2008;5(7):621–8 (Epub 2008/06/03).
89. Jiang H, Wong WH. Statistical inferences for isoform expression in RNA-Seq. Bioinformatics. 2009;25(8):1026–32 (Epub 2009/02/27).
90. Li B, Ruotti V, Stewart RM, Thomson JA, Dewey CN. RNA-Seq gene expression estimation with read mapping uncertainty. Bioinformatics. 2010;26(4):493–500 (Epub 2009/12/22).
91. Katz Y, Wang ET, Airoldi EM, Burge CB. Analysis and design of RNA sequencing experiments for identifying isoform regulation. Nat Methods. 2010;7(12):1009–15 (Epub 2010/11/09).
92. Consortium TE. Standards, Guideline and Best Practices for RNA-Seq. 2011; V1.0. Available from: https://www.encodeproject.org/.
93. Anders S, Huber W. Differential expression analysis for sequence count data. Genome Biol. 2010;11(10):R106 (Epub 2010/10/29).
94. Robinson MD, McCarthy DJ, Smyth GK. edgeR: a Bioconductor package for differential expression analysis of digital gene expression data. Bioinformatics. 2010;26(1):139–40 (Epub 2009/11/17).
95. Ritchie ME, Phipson B, Wu D, Hu Y, Law CW, Shi W, et al. limma powers differential expression analyses for RNA-sequencing and microarray studies. Nucleic Acids Res. 2015;43(7):e47 (Epub 2015/01/22).
96. da Huang W, Sherman BT, Lempicki RA. Systematic and integrative analysis of large gene lists using DAVID bioinformatics resources. Nat Protoc. 2009;4(1):44–57 (Epub 2009/01/10).
97. Liberzon A, Subramanian A, Pinchback R, Thorvaldsdottir H, Tamayo P, Mesirov JP. Molecular signatures database (MSigDB) 3.0. Bioinformatics. 2011;27(12):1739–40 (Epub 2011/05/07).

98. Hardcastle TJ, Kelly KA. baySeq: empirical Bayesian methods for identifying differential expression in sequence count data. BMC Bioinformatics. 2010;11:422 (Epub 2010/08/12).
99. Boyer LA, Lee TI, Cole MF, Johnstone SE, Levine SS, Zucker JP, et al. Core transcriptional regulatory circuitry in human embryonic stem cells. Cell. 2005;122(6):947–56.
100. Iyer VR, Horak CE, Scafe CS, Botstein D, Snyder M, Brown PO. Genomic binding sites of the yeast cell-cycle transcription factors SBF and MBF. Nature. 2001;409(6819):533–8.
101. Wu JQ, Seay M, Schulz, V., Hariharan, M., Tuck, D., Lian, J., Du, J., Shi, M., Ye, Z. J.,, Gerstein M, Snyder M, Weissman S. TCF7 is a key regulator of the self-renewal and differentiation switch in a multipotential hematopoietic cell line. PLoS Genetics. 2012;In Press.
102. Barski A, Cuddapah S, Cui K, Roh TY, Schones DE, Wang Z, et al. High-resolution profiling of histone methylations in the human genome. Cell. 2007;129(4):823–37 (Epub 2007/05/22).
103. Johnson DS, Mortazavi A, Myers RM, Wold B. Genome-wide mapping of in vivo protein-DNA interactions. Science. 2007;316(5830):1497–502 (Epub 2007/06/02).
104. Park PJ. ChIP-seq: advantages and challenges of a maturing technology. Nat Rev Genet. 2009;10(10):669–80 (Epub 2009/09/09).
105. Visel A, Blow MJ, Li Z, Zhang T, Akiyama JA, Holt A, et al. ChIP-seq accurately predicts tissue-specific activity of enhancers. Nature. 2009;457(7231):854–8 (Epub 2009/02/13).
106. Chen Y, Negre N, Li Q, Mieczkowska JO, Slattery M, Liu T, et al. Systematic evaluation of factors influencing ChIP-seq fidelity. Nat Methods. 2012;9(6):609–14 (Epub 2012/04/24).
107. Landt SG, Marinov GK, Kundaje A, Kheradpour P, Pauli F, Batzoglou S, et al. ChIP-seq guidelines and practices of the ENCODE and modENCODE consortia. Genome Res. 2012;22(9):1813–31 (Epub 2012/09/08).
108. Kharchenko PV, Tolstorukov MY, Park PJ. Design and analysis of ChIP-seq experiments for DNA-binding proteins. Nat Biotechnol. 2008;26(12):1351–9 (Epub 2008/11/26).
109. Li R, Li Y, Kristiansen K, Wang J. SOAP: short oligonucleotide alignment program. Bioinformatics. 2008;24(5):713–4 (Epub 2008/01/30).
110. Bailey T, Krajewski P, Ladunga I, Lefebvre C, Li Q, Liu T, et al. Practical guidelines for the comprehensive analysis of ChIP-seq data. PLoS computational biology. 2013;9(11): e1003326 (Epub 2013/11/19).
111. Jung LY, Kharchenko P, Wold B, Sidow A, Batzoglou S, Park P. Assessment of ChIP-seq data quality using cross-correlation analysis.
112. Muino JM, Kaufmann K, van Ham RC, Angenent GC, Krajewski P. ChIP-seq analysis in R (CSAR): an R package for the statistical detection of protein-bound genomic regions. Plant Methods. 2011;7:11 (Epub 2011/05/11).
113. Zhang Y, Liu T, Meyer CA, Eeckhoute J, Johnson DS, Bernstein BE, et al. Model-based analysis of ChIP-Seq (MACS). Genome Biol. 2008;9(9):R137 (Epub 2008/09/19).
114. Ji H, Jiang H, Ma W, Johnson DS, Myers RM, Wong WH. An integrated software system for analyzing ChIP-chip and ChIP-seq data. Nat Biotechnol. 2008;26(11):1293–300 (Epub 2008/11/04).
115. Qin ZS, Yu J, Shen J, Maher CA, Hu M, Kalyana-Sundaram S, et al. HPeak: an HMM-based algorithm for defining read-enriched regions in ChIP-Seq data. BMC Bioinf. 2010;11:369 (Epub 2010/07/06).
116. Bardet AF, He Q, Zeitlinger J, Stark A. A computational pipeline for comparative ChIP-seq analyses. Nat Protoc. 2012;7(1):45–61 (Epub 2011/12/20).
117. Tompa M, Li N, Bailey TL, Church GM, De Moor B, Eskin E, et al. Assessing computational tools for the discovery of transcription factor binding sites. Nat Biotechnol. 2005;23(1):137–44 (Epub 2005/01/08).
118. Ma W, Noble WS, Bailey TL. Motif-based analysis of large nucleotide data sets using MEME-ChIP. Nat Protoc. 2014;9(6):1428–50 (Epub 2014/05/24).

119. Rozowsky J, Euskirchen G, Auerbach RK, Zhang ZD, Gibson T, Bjornson R, et al. PeakSeq enables systematic scoring of ChIP-seq experiments relative to controls. Nat Biotechnol. 2009;27(1):66–75.
120. Valouev A, Johnson DS, Sundquist A, Medina C, Anton E, Batzoglou S, et al. Genome-wide analysis of transcription factor binding sites based on ChIP-Seq data. Nat Methods. 2008;5(9):829–34 (Epub 2009/01/23).
121. Fejes AP, Robertson G, Bilenky M, Varhol R, Bainbridge M, Jones SJ. FindPeaks 3.1: a tool for identifying areas of enrichment from massively parallel short-read sequencing technology. Bioinformatics. 2008;24(15):1729–30 (Epub 2008/07/05).
122. Boyle AP, Guinney J, Crawford GE, Furey TS. F-Seq: a feature density estimator for high-throughput sequence tags. Bioinformatics. 2008;24(21):2537–8 (Epub 2008/09/12).
123. Subramanian A, Tamayo P, Mootha VK, Mukherjee S, Ebert BL, Gillette MA, et al. Gene set enrichment analysis: a knowledge-based approach for interpreting genome-wide expression profiles. Proc Natl Acad Sci USA. 2005;102(43):15545–50.
124. Bryder D, Rossi DJ, Weissman IL. Hematopoietic stem cells: the paradigmatic tissue-specific stem cell. Am J Pathol. 2006;169(2):338–46.
125. Shizuru JA, Negrin RS, Weissman IL. Hematopoietic stem and progenitor cells: clinical and preclinical regeneration of the hematolymphoid system. Annu Rev Med. 2005;56:509–38.
126. Faubert A, Lessard J, Sauvageau G. Are genetic determinants of asymmetric stem cell division active in hematopoietic stem cells? Oncogene. 2004;23(43):7247–55.
127. Zhou JX, Huang S. Understanding gene circuits at cell-fate branch points for rational cell reprogramming. Trends Genet. 2011;27(2):55–62.
128. Waltzer L, Gobert V, Osman D, Haenlin M. Transcription factor interplay during Drosophila haematopoiesis. Int J Dev Biol. 2010;54(6–7):1107–15.
129. Bertrand V, Hobert O. Lineage programming: navigating through transient regulatory states via binary decisions. Curr Opin Genet Dev. 2010;20(4):362–8.
130. Jukam D, Desplan C. Binary fate decisions in differentiating neurons. Curr Opin Neurobiol. 2010;20(1):6–13.
131. Moore KA, Lemischka IR. "Tie-ing" down the hematopoietic niche. Cell. 2004;118(2):139–40.
132. Tsai S, Bartelmez S, Sitnicka E, Collins S. Lymphohematopoietic progenitors immortalized by a retroviral vector harboring a dominant-negative retinoic acid receptor can recapitulate lymphoid, myeloid, and erythroid development. Genes Dev. 1994;8(23):2831–41.
133. Pinto do OP. Kolterud A, Carlsson L. Expression of the LIM-homeobox gene LH2 generates immortalized steel factor dependent multipotent hematopoietic precursors. EMBO J. 1998;17(19):5744–56.
134. Yu WM, Hawley TS, Hawley RG, Qu CK. Immortalization of yolk sac-derived precursor cells. Blood. 2002;100(10):3828–31.
135. Sauvageau G, Iscove NN, Humphries RK. In vitro and in vivo expansion of hematopoietic stem cells. Oncogene. 2004;23(43):7223–32.
136. Ye ZJ, Kluger Y, Lian Z, Weissman SM. Two types of precursor cells in a multipotential hematopoietic cell line. Proc Natl Acad Sci USA. 2005;102(51):18461–6.
137. Raich N, Clegg CH, Grofti J, Romeo PH, Stamatoyannopoulos G. GATA1 and YY1 are developmental repressors of the human epsilon-globin gene. EMBO J. 1995;14(4):801–9.
138. Breitkreutz BJ, Stark C, Tyers M. Osprey: a network visualization system. Genome Biol. 2003;4(3):R22.
139. Horak CE, Luscombe NM, Qian J, Bertone P, Piccirrillo S, Gerstein M, et al. Complex transcriptional circuitry at the G1/S transition in *Saccharomyces cerevisiae*. Genes Dev. 2002;16(23):3017–33.
140. Lee TI, Rinaldi NJ, Robert F, Odom DT, Bar-Joseph Z, Gerber GK, et al. Transcriptional regulatory networks in *Saccharomyces cerevisiae*. Science. 2002;298(5594):799–804.

141. Harbison CT, Gordon DB, Lee TI, Rinaldi NJ, Macisaac KD, Danford TW, et al. Transcriptional regulatory code of a eukaryotic genome. Nature. 2004;431(7004):99–104.
142. Borneman AR, H. Yu, P. Bertone, M. Gerstein and M. Snyder. The transcription factors Mga1 and Phd1 are master regulators of a complex transcriptional network controlling pseudohyphal growth. Cell, submitted. 2005.
143. Weintraub H, Tapscott SJ, Davis RL, Thayer MJ, Adam MA, Lassar AB, et al. Activation of muscle-specific genes in pigment, nerve, fat, liver, and fibroblast cell lines by forced expression of MyoD. Proc Natl Acad Sci USA. 1989;86(14):5434–8.
144. Tapscott SJ. The circuitry of a master switch: Myod and the regulation of skeletal muscle gene transcription. Development. 2005;132(12):2685–95.
145. Asakura A, Lyons GE, Tapscott SJ. The regulation of MyoD gene expression: conserved elements mediate expression in embryonic axial muscle. Dev Biol. 1995;171(2):386–98.
146. Goldhamer DJ, Brunk BP, Faerman A, King A, Shani M, Emerson CP Jr. Embryonic activation of the myoD gene is regulated by a highly conserved distal control element. Development. 1995;121(3):637–49.
147. Kurokawa M, Hirai H. Role of AML1/Runx1 in the pathogenesis of hematological malignancies. Cancer Sci. 2003;94(10):841–6.
148. Friedman AD. Cell cycle and developmental control of hematopoiesis by Runx1. J Cell Physiol. 2009;219(3):520–4.
149. Coelho PS, Bryan AC, Kumar A, Shadel GS, Snyder M. A novel mitochondrial protein, Tar1p, is encoded on the antisense strand of the nuclear 25S rDNA. Genes Dev. 2002;16 (21):2755–60.
150. Tycowski KT, Shu MD, Steitz JA. A mammalian gene with introns instead of exons generating stable RNA products. Nature. 1996;379(6564):464–6.
151. Zhang Z, Harrison P, Gerstein M. Identification and analysis of over 2000 ribosomal protein pseudogenes in the human genome. Genome Res. 2002;12(10):1466–82.
152. Snyder M, Gerstein M. Genomics. Defining genes in the genomics era. Science. 2003;300 (5617):258–60.

Author Biographies

Dr. Raquel Cuevas Diaz Duran obtained her PhD degree in biotechnology from Tecnologico de Monterrey, Mexico, in 2014 working in the Cell Therapy Research Group. Her thesis involved the genomic characterization of adipose-derived stem cells and the identification of regulatory modules driving differentiation using time-series gene expression. In 2015, she joined Prof. Jiaqian Wu's laboratory in the Vivian L. Smith Department of Neurosurgery at the University of Texas as a postdoctoral research fellow. Her research is focused on the implementation of systems-based approaches for the integration of epigenetic, transcriptomic, and proteomic data obtained from time-series induction experiments of stem cells in order to identify molecular mechanisms regulating fate commitment and differentiation.

Dr. Sudheer Menon obtained his PhD degree in bioinformatics in the Department of Bioinformatics, Bharathiar University and DBEB, IIT Delhi, India, in 2010. His thesis was on "Computational identification of promoter regions in fungal genome." He worked as a postdoctoral fellow in Bioinformatics at Umea University, Sweden, and University of Witwatersrand, Johannesburg, in 2011–2012. Subsequently, he joined as a postdoctoral visiting fellow at National Institutes of Health, Bethesda, Maryland, in 2013 where his study mainly focused on "Characterization of transcriptional end sites and gene looping in mouse and human genome." In 2015, he worked in Prof. Jiaqian Wu's laboratory at the University of Texas as a postdoctoral research fellow. He has presented his research at several national and international conferences.

Prof. Jiaqian Wu An assistant professor in the Vivian L. Smith Department of Neurosurgery and Center for Stem Cell and Regenerative Medicine at the University of Texas Medical School at Houston, and Dr. Wu earned her doctorate in molecular and human genetics at Baylor College of Medicine and did her postdoctoral work at Yale and Stanford University. The Wu laboratory combines stem cell biology and systems-based approaches involving functional genomics, bioinformatics, and NGS technologies to unravel gene transcription and regulatory mechanisms governing neural and blood development and differentiation. Dr. Wu's work has been recognized with prestigious honors and awards, including the National Institute of Health Pathway to Independence (PI) Award (K99/R00), R01 and the Senator Lloyd & B.A. Bentsen Investigator Award which she currently holds; the National Institutes of Health Ruth L. Kirschstein National Research Service Award for Individual Postdoctoral Fellows; and the International Society for Stem Cell Research (ISSCR) Annual Meeting Travel Award. A reviewer for NIH, MRC, the journals Nucleic Acids Research, Genome Research, and Genome Biology, Dr. Wu has presented invited talks and lectures on stem cell biology, functional genomics, and proteomics at international conferences, the Multiple Sclerosis Research Center of New York, Lawrence Livermore National Laboratory, the University of Florida, etc. She has developed a patent, authored a book, and wrote many articles that have appeared in PNAS, Genome Biology, Plos Genetics, Genome Research, the Journal of Neuroscience, and Nature.

Chapter 2
Global Approaches to Alternative Splicing and Its Regulation—Recent Advances and Open Questions

Yun-Hua Esther Hsiao, Ashley A. Cass, Jae Hoon Bahn, Xianzhi Lin and Xinshu Xiao

Abstract Pre-mRNA splicing is an essential RNA processing step in eukaryotes. Alternative splicing generates distinct spliced isoforms of the same gene, thereby dramatically increasing transcriptome diversity. Since most human genes undergo alternative splicing, this process contributes to a wide spectrum of biological functions in healthy and disease states. Splicing is closely regulated by various *cis*-regulatory elements and *trans*-factors. With the advent of high-throughput experimental technologies and bioinformatic algorithms, we now have powerful means to study alternative splicing globally and uncover its functional impact and regulatory mechanisms. As more RNA sequencing (RNA-Seq) data from normal and disease conditions are becoming available, many studies are underway to dissect global misregulation of splicing in diseases and develop novel splicing-targeted therapeutics. In this chapter, we first discuss the experimental and bioinformatic approaches for identification of alternative splicing, followed by a comprehensive review on the state-of-the-art methodologies to study splicing regulation. In addition, we discuss the current challenges and open questions in the RNA splicing field

Y.-H.E. Hsiao · X. Xiao (✉)
Department of Bioengineering, University of California Los Angeles, Los Angeles, CA 90095, USA
e-mail: gxxiao@ucla.edu

A.A. Cass · X. Xiao
Bioinformatics Interdepartmental Program, University of California Los Angeles, Los Angeles, CA 90095, USA

J.H. Bahn · X. Lin · X. Xiao
Department of Integrative Biology and Physiology, University of California Los Angeles, Los Angeles, CA 90095, USA

X. Xiao
Molecular Biology Institute, University of California Los Angeles, Los Angeles, CA 90095, USA

X. Xiao
UCLA, Terasaki Life Sciences Building 2000A, 610 Charles E. Young Drive, Los Angeles, CA 90095-1570, USA

J. Wu, *Transcriptomics and Gene Regulation*,
Translational Bioinformatics 9, DOI 10.1007/978-94-017-7450-5_2

including gene expression kinetics, co-transcriptional splicing, and therapeutic approaches targeting splicing.

Keywords Alternative splicing · RNA · RNA-Seq · Gene regulation

2.1 Introduction

First discovered nearly 40 years ago [1, 2], pre-mRNA splicing consists of a series of biochemical reactions that function to remove introns and ligate flanking exons. Exon–intron boundaries are defined by highly conserved consensus sequences including the 5′ splice site (5′ss, or donor site), 3′ splice site (3′ss, or acceptor site), and branch point sequences (BPSs) (Fig. 2.1). These sequences are recognized by the spliceosome, a dynamic multi-ribonucleoprotein complex composed of small nuclear ribonucleoproteins (snRNPs) (refer to [3] for detailed reviews). The spliceosome is the basic machinery that carries out splicing reactions.

In recent years, it was estimated that more than 90 % of human genes are processed through alternative splicing where multiple spliced isoforms are generated from a single gene, thus significantly increasing transcriptome diversity [4–6]. The most extreme case of alternative splicing is the *Drosophila* Down Syndrome cell adhesion molecule gene (*Dscam*) which includes 48 exons and can theoretically produce 38,016 alternative transcripts from a single gene [7]. Different types of alternative splicing exist with the most common ones being exon skipping, alternative 5′ss usage, alternative 3′ss usage, mutually exclusive exons, and intron retention [8].

It is now well established that alternative splicing contributes to a wide spectrum of cellular functions [9]. Disruption of normal splicing was reported for a large number of human diseases, which has been reviewed extensively [10–12]. As a functionally critical process, alternative splicing is regulated by a myriad of *cis*-elements and *trans*-acting factors (Fig. 2.1). Splicing regulatory elements (SREs) reside in exons or introns and function to either enhance or silence splicing. These *cis*-elements are thus named accordingly as: exonic splicing enhancers (ESEs), intronic splicing enhancers (ISEs), exonic splicing silencers (ESSs), and intronic splicing silencers (ISSs). These *cis*-elements interact with many *trans*-acting factors (i.e., splicing factors), including serine/arginine-rich (SR) proteins and heterogeneous nuclear ribonucleoproteins (hnRNPs) [13]. RNA secondary structures also affect alternative splicing, likely by facilitating or blocking accessibility of splicing factors to their cognate RNA [14].

Understanding the regulatory mechanisms of alternative splicing in health and disease is an essential topic of gene regulation. Recent advances in high-throughput technologies and related bioinformatic methodologies are enabling exciting discoveries in this area. Here, we first focus on global approaches for splicing identification, followed by an in-depth review of methodologies to study splicing regulatory mechanisms.

2.2 Identification and Validation of Alternative Splicing Events

2.2.1 Identification of Alternative Splicing Events

2.2.1.1 High-Throughput Experimental Approaches

The first high-throughput method developed to detect and quantify alternative splicing events was customized microarrays [15–17]. An initial study by Hu et al. [18] used multi-probe design of Affymetrix arrays to detect splicing variants, demonstrating the utility of microarrays for splicing analyses. Later studies [19, 20] developed different techniques to improve the microarray probe design and successfully profiled alternative splicing events and their expression on the genome-wide scale. Johnson et al. [19] used splice junction arrays to probe around 10,000 human multi-exon genes across 52 tissues. Besides the known alternative splicing events, they were also able to discover novel spliced isoforms of many genes. Pan et al. [20] took the focused probe design approach (see review [16]) with three exon body probes and three spliced junction probes for each known alternative splicing event to achieve more sensitive expression quantification. In this study, they were able to globally determine the tissue specificity of alternative splicing events in mouse tissues. Many recent studies adopted different probe designs and microarray platforms to investigate splicing profiles and splicing levels in healthy and disease samples (reviewed in [15–17]).

Since the advent of next-generation sequencing (NGS), RNA-Seq became an essential technology for global studies of alternative splicing (Fig. 2.2a). It provides a means to directly or indirectly sequence the RNA molecules in a high-throughput manner. At present, often-used RNA-Seq methods first convert the RNA sample of interest into cDNAs, which are then made into a sequencing library that consists of short DNA fragments (corresponding to the RNA of interest) flanked by pre-designed adapter oligos. The DNA library is then sequenced from one end (single-end sequencing, or SE) or both ends (paired-end sequencing, or PE) to yield final RNA-Seq reads [21]. The resulting RNA-Seq reads correspond to a snapshot of RNA expression in the respective cellular sample.

RNA-Seq is advantageous in several ways. First, it can detect novel isoforms and alternative splicing events that are not yet annotated [22, 23]. Second, RNA-Seq is not affected by the cross-hybridization problem that confounds many microarray-based studies [21]. Third, RNA-Seq data can provide relatively accurate quantification of levels of gene expression and splicing [21, 24]. Lastly, RNA-Seq provides single-nucleotide information that enables studies of genetic variants [25, 26] and RNA editing sites [27–30], in addition to gene or exon expression. Using RNA-Seq, a large number of alternative splicing events were identified in human and mouse tissues [4, 5].

Although RNA-Seq has dramatically improved our knowledge on alternative splicing, there are still remaining challenges to be addressed [21, 31]. RNA-Seq

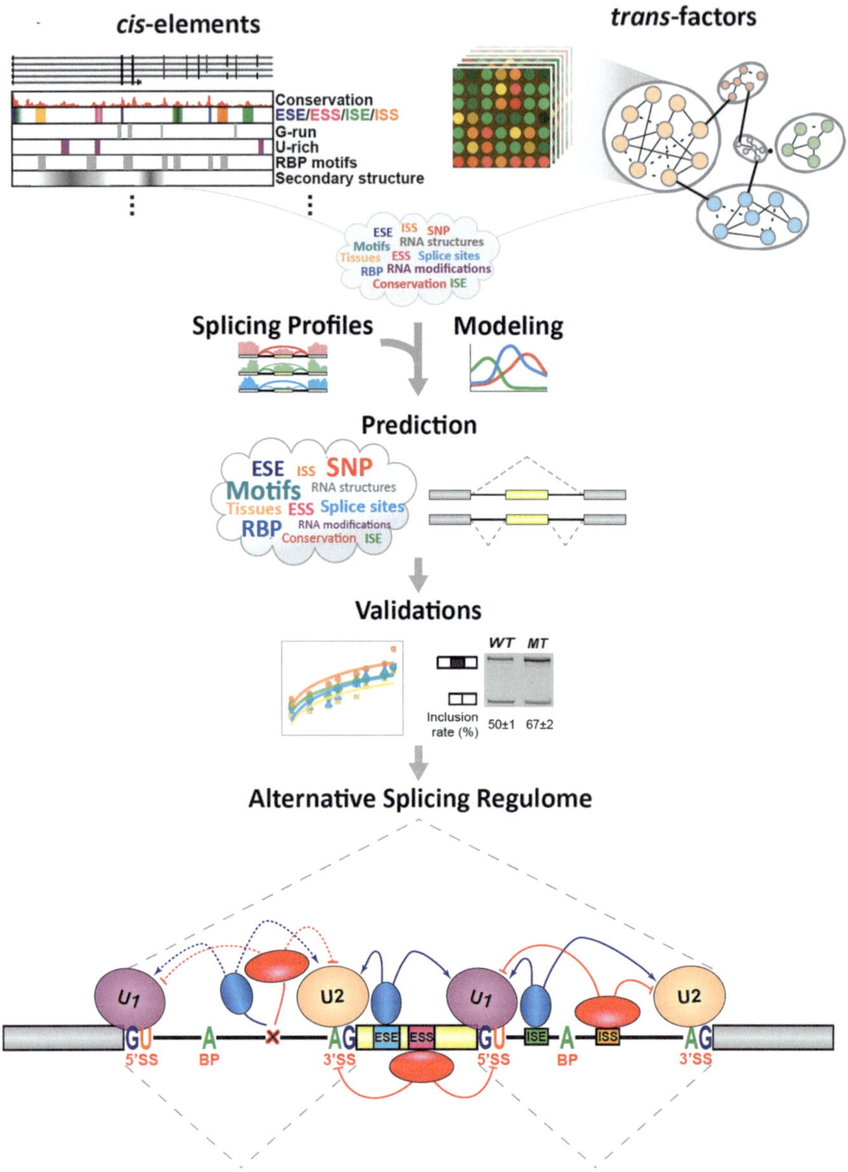
cis-elements
trans-factors
Conservation
ESE/ESS/ISE/ISS
G-run
U-rich
RBP motifs
Secondary structure
ESE ISS SNP Motifs RNA structures Tissues ESS Splice sites RBP RNA modifications Conservation ISE
Splicing Profiles
Modeling
Prediction
ESE ISS SNP Motifs RNA structures Tissues ESS Splice sites RBP RNA modifications Conservation ISE
Validations
WT MT
Inclusion rate (%)
50±1 67±2
Alternative Splicing Regulome
U1
U2
U1
U2
GU
A
AG
ESE
ESS
GU
ISE
A
ISS
AG
5'SS
BP
3'SS
5'SS
BP
3'SS

◀ **Fig. 2.1** Overview of previous studies in alternative splicing regulation. *Cis*-regulatory elements and *trans*-acting factors are key components in the splicing regulatory networks (alternative splicing regulome), which have been actively examined. Combined with global profiles of alternative splicing patterns, bioinformatic models were developed to predict the relative impacts of different regulators and splicing outcome of a given exon. Experimental validations are critical steps to evaluate the accuracy of the predicted splicing regulation. The *bottom* diagram illustrates well-known components of splicing regulation. The *yellow box* represents an alternatively skipped exon, which has ESE and ESS motifs that can be recognized by splicing factors. The flanking introns of this exon harbor ISE and ISS motifs. Interactions between the splicing factors and the core splicing machinery (U1, U2 snRNPs, etc.) are illustrated. Splicing enhancers (ESEs, ISEs) normally promote exon inclusion, which is represented by the *arcs* with *arrowheads*, whereas splicing silencers (ESSs, ISSs) repress exon inclusion, which is represented by *flat-headed arcs*. Genetic variants may disrupt splicing motifs and alter the binding strength of splicing factors (illustrated by the *x*). Other mechanisms such as RNA modifications or RNA secondary structures may also affect alternative splicing, which are not illustrated in this diagram. ESE: exonic splicing enhancer; ESS: exonic splicing silencer; ISE: intronic splicing enhancer; ISS: intronic splicing silencer; 5'ss: 5' splice site; 3'ss: 3' splice site and BP: branch point

library construction is the first critical step. Different library preparation protocols were developed to study various biological questions and thus have their own merits and limitations [32, 33]. In addition, RNA-Seq library generation protocols often need optimization for specific RNA samples based on sample quality, concentration, and other variables. In large-scale experiments, batch effects in RNA-Seq data may be a critical problem to consider [34], which may mislead study conclusions if not properly accounted for. Finally, RNA-Seq experiments are still costly, especially for studies of alternative splicing. In such applications, reads covering spliced junctions are examined closely to guide the identification and quantification of alternative splicing. Thus, it is highly desirable to have a relatively large number of spliced reads. Often-used settings of RNA-Seq in splicing studies favor PE reads, long read length (e.g., >75 bp), and high sequencing depth (≥100 million PE reads for human samples) [35, 36].

Alternative approaches were developed to address some of the above challenges in RNA-Seq. For example, RNA-mediated oligonucleotide annealing, selection, and ligation with next-generation sequencing (RASL-Seq) allows for RNA-Seq of a limited set of exons in hundreds or thousands of biological samples [37] (Fig. 2.2b). Thus, it is ideal for large-scale analysis of up to 500 exons in complex networks or pathways [37]. The main difference between RNA-Seq and RASL-Seq is the use of oligonucleotides that recognize a specific spliced junction in the latter method. After ligating the pairs of oligos, these specific RNAs are then isolated with biotinylated oligo-dTs and pulled down with streptavidin-coated magnetic beads. A unique barcode for each sample is incorporated during PCR, allowing for pooled sequencing of >1500 samples per lane [37]. Analyzing expression of a limited number of genes in many samples has clinical applications, such as screening for drugs that inhibit splicing events implicated in cancer [38]. One factor of consideration in RASL-Seq is the efficiency and specificity of ligation; Rnl2 was shown to have higher efficiency than T4 ligase [39].

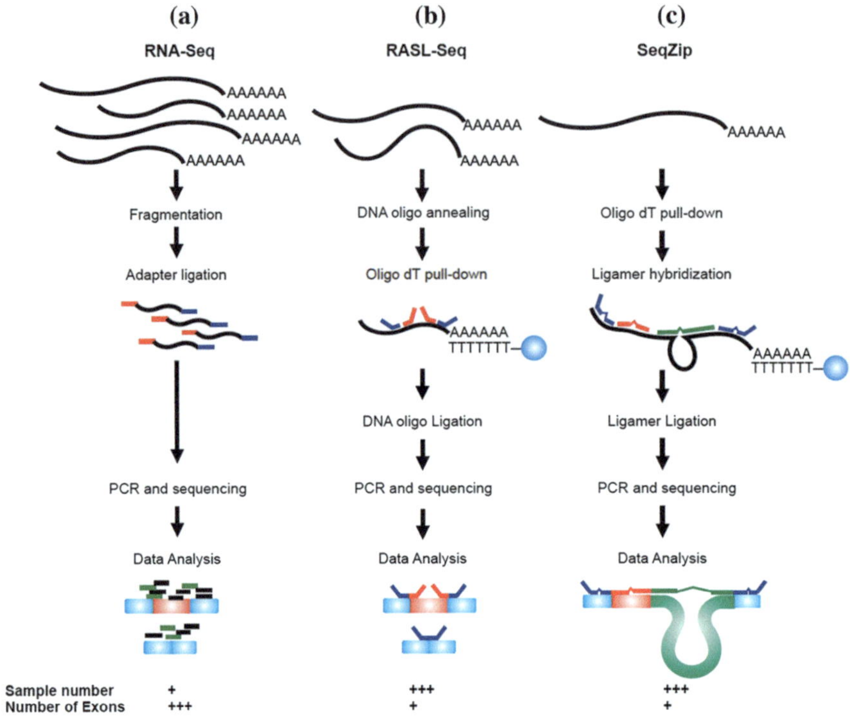

Fig. 2.2 High-throughput experimental approaches for splicing detection. **a** RNA-Seq, the most popular method for splicing analysis, begins with creating cDNA libraries of fragmented RNA. Then, sequencing adapters are added to make a sequencing library, followed by PCR amplification and sequencing. In data analysis, reads that span spliced exon junctions and those that are located within exon bodies are identified bioinformatically to detect and quantify alternative splicing. This method can provide data for many expressed exons (+++) in the sample of interest. The cost of RNA-Seq is relatively high, which may limit the number of samples (+) that can be analyzed in a specific study. **b** RASL-Seq requires a pair of pre-designed oligonucleotides that recognize specific splice junctions of intact (i.e., unfragmented) mRNA. Biotinylated oligo-dTs with streptavidin-coated magnetic beads are then used to pull down the RNA. Barcode incorporation during PCR allows for pooled sequencing of ~1500 samples per sequencing lane. Compared to RNA-seq, RASL-Seq is ideal for few (up to 500) exons (+) in hundreds or thousands of samples (+++). **c** SeqZip uses DNA "ligamers" to directly sequence long transcript isoforms, causing intermediate regions to loop out. Compared to RNA-Seq and RASL-Seq, SeqZip is specialized for targeting long transcripts

A limitation common to all sequencing-based methods is the sequencing read length, which is typically much shorter than the full-length isoform of long transcripts. Full-length isoforms are thus reconstructed computationally using overlapping reads, though there is always a degree of uncertainty [40, 41]. To overcome this limitation, a new method SeqZip was recently developed [42] (Fig. 2.2c). It uses ~40–60nt DNA "ligamers" that recognize the 5′ and 3′ ends of single or multiple alternatively spliced exons that may be thousands of nucleotides apart, causing the intermediate sequence to loop out [42]. Multiple ligamers hybridized to the same transcript are then ligated together, thereby connecting distant exons in the same transcript. Assessing the length and sequence of the DNA ligamers allows for deduction of the full-length isoform. Thus, SeqZip greatly improves the ability to sequence long transcripts.

Aside from the above-mentioned RNA-based approaches, protein-based approaches may be used to identify changes in protein expression resulting from alternative splicing events. Mass spectrometry has been used to identify alternative splicing events in breast cancer [43]. Still, RNA-based approaches are far more commonly used for alternative splicing identification. The choice of experimental method depends on the experimental goal. As sequencing technology improves, so will the ability to identify alternative splicing events.

2.2.1.2 Bioinformatic Algorithms for Analyzing Alternative Splicing Using RNA-Seq

Current bioinformatic methods for analyzing alternative splicing in RNA-Seq can be largely classified into two categories: exon-centric and isoform-centric. Exon-centric approaches directly estimate the splicing level of each exon typically by calculating its percent spliced-in (PSI) [4], a measure of the frequency of exon inclusion among all mature mRNA molecules of the gene (also see reviews [44, 45]). In contrast, isoform-centric methods aim to quantify the abundance of each alternative isoform of the gene, which can be followed by further comparisons to determine differential splicing [46–48].

The benefit of using exon-centric splicing detection is that the type and PSI of each alternatively spliced exon are directly interrogated. Such single-exon information is useful in designing experiments to validate and further examine these events [36, 49]. PSI can be calculated in different ways. First, abundance of reads aligned directly to alternative exon junctions is used, with the exon body reads optionally included [36, 49]. However, it is difficult to precisely estimate the PSI value in cases of complex alternative splicing. To overcome this problem, other tools, such as SplAdder and DiffSplice [50, 51], adopt a splicing graph strategy to capture the complexity of alternative splicing by building a graph of spliced isoforms where nodes represent exons and edges represent spliced introns. Input RNA-Seq data are used to update the alternative path in the graph. The challenge in these approaches is that the splicing graph can be complicated by poorly supported events, so post-filtering is necessary to reduce false positives. In general, exon-centric methods alone do not support identification of novel alternative splicing events due to their requirement of gene annotation.

Instead of focusing on specific splicing events, isoform-centric methods use RNA-Seq to construct isoforms and estimate their expression levels [52–55]. Most tools also utilize the reference genome to guide isoform reconstruction, but others perform de novo transcriptome assembly without relying on the reference genome. The latter type is particularly helpful for alternative splicing analyses in species with poorly annotated genomes. Early isoform-centric methods were developed under the assumption that the read distribution is uniform, though this is rarely the case. New methods are now available to account for RNA-Seq read non-uniformity [56, 57]. Another recent development for isoform-centric analysis is the alignment-free approach, which bypasses the time-consuming alignment step by building a hash index from the reference transcripts using sequence k-mers as keys and applying an expectation maximization algorithm to estimate isoform abundance [46, 47]. This approach speeds up the computational time considerably while maintaining prediction accuracy. However, it remains to be evaluated whether such methods perform well in the presence of sample-specific genetic variants.

Once alternative splicing is identified, both classes of methods provide a means to detect differentially spliced events. The outcome from exon-centric analyses is a list of differentially spliced events that can be directly used for further analysis (e.g., experimental validation, functional interpretation, and regulatory studies). On the other hand, isoform-centric analysis captures the splicing complexity of a series of related events within the same isoform, but further steps are often needed to pinpoint individual splicing events of interest. In Table 2.1, we summarize often-used tools for splicing analysis.

2.2.2 Validation of Alternative Splicing Events

In silico tools that detect alternative splicing events based on RNA-Seq data usually generate a large number of candidates. A subset of these events should be experimentally validated in vivo or in vitro. Verification experiments for alternative splicing events are readily carried out by reverse transcription followed by PCR (RT-PCR) using primers that target flanking constitutive exons [58]. This strategy works well for alternative splicing events in genes with intermediate or high expression levels. In order to verify lowly expressed events, in vitro minigene expression analysis by RT-PCR can be utilized [59–61]. Compared with in vivo assays, the minigene system is able to validate events regardless of their endogenous expression level. However, since only a limited region flanking the exon of interest can be cloned into the minigene vector, this in vitro approach may not faithfully reproduce in vivo splicing patterns. It should be noted that both types of experiments are considered low-throughput and labor intensive, thus only validating a relatively small number of events.

High-throughput methods for validation of alternative splicing events are in great demand and several such approaches are on the horizon. For example, RT-PCR experiments may be scaled up when used in conjunction with microfluidic devices [62]. In addition, recent methods, such as the "designer exons" approach [63], may be further developed for this purpose. With the rapid technology development in synthetic biology and genome editing, it is likely that high-throughput splicing validation will soon become a reality.

Table 2.1 Tools for analysis of alternative splicing using RNA-Seq data

Function	Category	Program	URL	Input	Gene annotation
AS prediction	Isoform-centric	Cufflinks	http://cole-trapnell-lab.github.io/cufflinks/cufflinks	SAM or BAM files	No
AS prediction	Isoform-centric	eXpress	http://bio.math.berkeley.edu/eXpress	SAM or BAM files	No
AS prediction	Isoform-centric	Trinity	http://trinityrnaseq.github.io	FASTQ files	Yes
AS prediction	Isoform-centric	Trans-ABySS	http://www.bcgsc.ca/platform/bioinfo/software/trans-abyss	FASTQ files	Yes
AS prediction	Isoform-centric	Scripture	http://www.broadinstitute.org/software/scripture	SAM or BAM files	No
AS prediction	Isoform-centric	RSEM	http://deweylab.biostat.wisc.edu/rsem	FASTA or FASTQ files, transcript expression	Yes
AS prediction	Isoform-centric	PennSeq	http://sourceforge.net/projects/pennseq	UCSC genome browser gene annotation, SAM files	Yes
AS prediction	Isoform-centric	RNA-Skim	http://www.csbio.unc.edu/rs	Specialized transcriptome FASTA, RNA-Seq FASTQ files	Yes
AS prediction	Isoform-centric	Sailfish	http://www.cs.cmu.edu/~ckingsf/software/sailfish	*k*-mer size, RNA-Seq in FASTA or FASTQ, transcriptome in GTF	Yes
AS prediction	Isoform-centric	Sequgio	http://fafner.meb.ki.se/biostatwiki/sequgio	BAM files	Yes
AS prediction	Exon-centric	SplAdder	https://github.com/ratschlab/spladder	GFF annotation, BAM files	Yes
AS prediction	Exon-centric	SpliceTrap	http://rulai.cshl.edu/splicetrap/doc/help.html	Gene annotation in BED or GTF format, TXdb exon isoform database, FASTA or FASTQ files	Yes
AS prediction	Exon-centric	ESFinder	http://mlg.hit.edu.cn/ybai/ES/ESFinder.html	GTF annotation, UCSC genome browser AS events, BAM files	Yes
AS prediction	Exon-centric	SplicePie	https://github.com/pulyakhina/splicing_analysis_pipeline	GTF annotation, BAM and BAM index files	Yes
AS prediction	Both	MISO	https://miso.readthedocs.org/en/fastmiso	GFF annotation, BAM files	Yes

(continued)

Table 2.1 (continued)

Function	Category	Program	URL	Input	Gene annotation
DAS	Isoform-centric	CuffDiff 2	http://cole-trapnell-lab.github.io/cufflinks/cuffdiff/index.html	GFF/GTF annotation, SAM files	Yes
DAS	Isoform-centric	IUTA	http://www.niehs.nih.gov/research/resources/software/biostatistics/iuta/index.cfm	GFF annotation, BAM files	Yes
DAS	Isoform-centric	SplicingCompass	http://www.ichip.de/software/SplicingCompass.html	Read coverage in GFF, BED files	Yes
DAS	Isoform-centric	rSeqDiff	http://www-personal.umich.edu/~jianghui/rseqdiff	"sampling_rates" files from rSeq [167], gene expression files	Yes
DAS	Isoform-centric	FDM	http://csbio-linux001.cs.unc.edu/nextgen/software/FDM	GTF annotation, SAM files	Yes
DAS	Isoform-centric	rDiff	http://bioweb.me/rdiff	GFF annotation, BAM files	Yes
DAS	Exon-centric	DEXSeq	http://www.bioconductor.org/packages/release/bioc/html/DEXSeq.html	GFF/GTF annotation, SAM files	Yes
DAS	Exon-centric	DSGSeq	http://bioinfo.au.tsinghua.edu.cn/software/DSGseq	BAM and BED files	Yes
DAS	Exon-centric	DiffSplice	http://www.netlab.uky.edu/p/bioinfo/DiffSplice	Parsed SAM files, program configuration files	No
DAS	Exon-centric	dSpliceType	http://dsplicetype.sourceforge.net	GFF annotation, read coverage in bedgraph format, junction files in BED format	Yes
DAS	Exon-centric	MATS/rMATS	http://rnaseq-mats.sourceforge.net	GFF annotation, FASTQ or BAM files, Bowtie index	Yes
DAS	Exon-centric	rSeqNP	http://www-personal.umich.edu/~jianghui/rseqnp	Transcript expression files	No
DAS	Both	MISO	https://miso.readthedocs.org/en/fastmiso	GFF annotation, BAM files	Yes

(continued)

Table 2.1 (continued)

Function	Category	Program	URL	Input	Gene annotation
DAS	Both	SUPPA	https://bitbucket.org/regulatorygenomicsupf/suppa	GTF annotation, transcript expression files	Yes
Spliced mapper	*NA*	HMMSplicer	http://derisilab.ucsf.edu/software/hmmsplicer	FASTQ files, Bowtie index	Yes
Spliced mapper	*NA*	PASSion	https://trac.nbic.nl/passion	FASTQ files, SMALT index	Yes
Spliced mapper	*NA*	PASTA	http://www.biotech.ufl.edu/cores/bioinformatics/dibig/dibig-software/pasta	FASTQ files, Bowtie index	Yes
Spliced mapper	*NA*	OLego	http://zhanglab.c2b2.columbia.edu/index.php/OLego	FASTA or FASTQ files, BWT index	Yes
Spliced mapper	*NA*	TrueSight	http://bioen-compbio.bioen.illinois.edu/TrueSight/	FASTQ files, Bowtie index	Yes
Spliced mapper	*NA*	UnSplicer	http://opal.biology.gatech.edu/paul/unsplicer/index.htm	FASTQ files, Bowtie index	Yes
Spliced mapper	*NA*	JAGuaR	http://www.bcgsc.ca/platform/bioinfo/software/jaguar	FASTQ files, BWA index	Yes
Spliced mapper	*NA*	Rail-RNA	https://github.com/nellore/rail	FASTQ files, Bowtie index	Yes

AS: alternative splicing; DAS: differential alternative splicing; Spliced mapper: read aligners specialized in mapping junction reads

2.3 Methodologies for Studies of Splicing Regulation

Pre-mRNA splicing is regulated by a large number of *cis*-elements and *trans*-acting factors. In this section, we will review the bioinformatic and experimental approaches for the identification and analysis of splicing regulatory mechanisms.

2.3.1 Cis-Regulation of Alternative Splicing

2.3.1.1 Splice Site Consensus Sequence

Splice site sequences are among the best-characterized *cis*-elements in splicing regulation, owing to the simplicity of their identification. Each internal exon is flanked by a 5′ss and a 3′ss. Thus, splice site sequences can be easily collected based on gene annotation. The majority of human exons are flanked by the GU-AG canonical sequences. However, the splice site signals normally involve a much longer sequence motif, which confers specificity and a dynamic range of splice site strength. Using known splice site sequences as training data, many algorithms were developed to predict splice site strength (see reviews [64, 65]). The most intuitive model is the position weight matrix (PWM), which is straightforward to implement but fails to consider the positional dependency between nucleotides in the splice site [66]. Other algorithms adopt more sophisticated probabilistic models such as neural networks or maximum entropy to more accurately estimate the splice site scores [67, 68].

2.3.1.2 Branch Point Sequences (BPSs)

The prediction of BPS is challenging because its location in the intron can be highly variable. For example, a BPS may be close to the 3′ss ($\sim$40nt upstream) or 100–400nt upstream of the 3′ss in the AG exclusion zone (AGEZ) [69]. Additionally, the BPS motif is highly degenerate [70] and multiple potentially functional BPSs may exist in a particular intron. A number of bioinformatic methods were developed to identify BPS and evaluate their strength. Human Splice Finder [66] uses PWMs and the algorithm proposed by Gooding et al. [69] to search for BPS candidates in a limited region. Another predictive approach makes use of sequence conservation and partial sequence complementarity of U2 snRNA to the BPS [71, 72]. A recent study showed that using machine learning methods such as support vector machines together with polypyrimidine and other sequence information could increase accuracy in BPS prediction [73]. Pastuszak et al. took advantage of the fact that Splicing Factor 1 (SF1) recognizes BPSs and restricted their motif analysis to sites with high SF1 binding affinity to predict BPS with relatively high accuracy [74].

Recently, a few studies used the NGS technology to identify BPS globally. In the RNA-Seq data, a minority of reads may derive from the junction of the 5′ss-branch point of the intron lariat. A search for such reads has led to successful identification of hundreds of BPS in human RNA-Seq data sets [75, 76]. The advantage of these approaches is that they do not require prior knowledge about the BPS locations or sequences. However, one drawback is that lariat reads are very rare among those generated from standard RNA-Seq libraries. Thus, very deeply sequenced data sets are needed to obtain adequate lariat read coverage. Another NGS-based method, called CaptureSeq [77], was applied recently to identify BPS [78]. In this method, tiling arrays were designed that contain oligonucleotide probes to target the 5′ss-branch point junctions [78]. cDNAs from the RNA samples of interest were then hybridized, eluted, and sequenced. As a complementary approach, RNase R digestion was applied to enrich for reads containing BPS without requiring pre-designed arrays. This study identified >50,000 human BPS in >10,000 genes, which enabled further investigation of global features of this class of splicing regulatory signal [78].

2.3.1.3 Splicing Regulatory Elements

Besides the core splicing signals, a large number of motifs in the exons or introns can also regulate splicing (Fig. 2.1) [8]. Identification and characterization of these SREs are instrumental to the understanding of splicing regulatory mechanisms. In general, genome-wide experimental or bioinformatic screens have been designed to identify SREs. Wang et al. developed the first large-scale screen of ESSs using splicing reporter assays in cultured cells [59]. This effort successfully identified hundreds of ESS sequences and shed light on the global properties of these elements. Later, a number of other experimental screens were carried out to identify different types of SREs [79–82]. These studies greatly expanded the catalog of known or predicted SREs without the associated *trans*-factors necessarily identified. Other experimental methods that pinpoint SREs for known splicing factors will be discussed later.

In addition to the experimental approaches, bioinformatic methods are also essential to SRE studies. Fairbrother et al. developed a motif comparison approach, RESCUE-ESE, to identify ESEs by evaluating motif enrichment correlated with different features of splicing [83]. Similar principles were applied later to identify other types of SREs [84, 85]. A myriad of other bioinformatic methods were also developed for this purpose, such as those based on comparative genomics [86], PWMs [87], or machine learning techniques [88–91].

With the increasing number of SREs, a great deal of effort was dedicated to understand the functional interaction among different elements and their context-dependent roles in splicing regulation. For example, Bayesian networks were used to study coevolutionary relationships of SREs in eukaryotes that reflect functional interaction [60]. Bioinformatic and statistical methods, combined with experimental approaches, were used to infer combinatorial function of different types of SREs [92–94]. The function of individual motifs (corresponding to one splicing factor) was studied in detail via bioinformatic modeling and analysis to reveal their context-dependent function globally [61, 95–97] (refer to [98] for a detailed review of this topic).

2.3.2 Genetic Variants Associated with Splicing

Genetic variants [such as mutations or single-nucleotide polymorphisms (SNPs)] play important roles in gene regulation because they can potentially alter *cis*-regulatory motifs. Previous studies estimated that 15–60 % of point mutations that result in human genetic diseases disrupt splicing [10, 99–102]. In recent years, exciting progress has been made in analyzing the involvement of genetic variants in modulating alternative splicing, which is reviewed in this section.

2.3.2.1 Splicing QTLs

Splicing quantitative trait loci (sQTL) analysis is an often-used method to identify SNPs associated with splicing phenotypes. In this method, the correlation between SNP genotypes and exon inclusion levels is examined using different means, ranging from simple linear correlation to model-based analysis [103–105]. Early sQTL studies used microarrays to detect isoform or exon expression levels, which is rapidly replaced by RNA-Seq-based analysis. However, this method requires a large number of samples to achieve adequate statistical power. In addition, sQTL analyses only deduce correlative relationships, without the capability of pinpointing the causal SNP for splicing alteration.

2.3.2.2 Machine Learning-Based Methods

In contrast to sQTLs, methods based on machine learning principles aim to predict the functional (causal) SNP that modulates alternative splicing. Different types of machine learning or statistical methods were adopted for this purpose [106–108]. One study used a random forest-based strategy and predicted exonic splicing-altering variants [106]. Another study developed a splicing code where "code quality" was optimized using information theory on a large number of features [109]. This splicing code was applied to predict genetic variants that may alter splicing [108–110]. One common challenge to such approaches is the limited availability of training data sets that should include experimentally validated SNPs with confirmed function in splicing and those that are known to have no influence on splicing. To overcome this problem, previous studies used disease-causing exonic mutations from existing databases as positive training data set and common SNPs in the general population as negative data set (assuming they do not affect splicing) [106, 107]. In contrast, the splicing code-based studies used human RNA-Seq data of different tissues to derive the code, without the need of direct model training using splicing-related variants [108–110].

2.3.2.3 Allele-Specific Alternative Splicing

To infer genetic regulation of alternative splicing, another powerful approach is built upon allele-specific expression (ASE) of genetic variants. ASE refers to the biased expression of the two alleles of a variant in diploid cells. RNA-Seq data provide single-nucleotide information that is appropriate for ASE studies. One advantage of ASE analysis is that the two alleles of a variant serve as within-sample controls of each other, which naturally eliminates the environmental and *trans*-acting effects that might alter splicing patterns or introduce variance in the data [111]. Nevertheless, one challenge in using RNA-Seq for ASE analysis lies in the step of read mapping. It is now clear that standard mapping methods induce a mapping bias that favors the reference allele of the genetic variant because the reference genome is utilized in mapping [112, 113]. Various strategies were developed to reduce this type of bias [27, 28]. Once ASE patterns are identified, they can be further analyzed to detect allele-specific alternative splicing events, as proposed in [25]. While sQTL studies and machine learning methods necessitate many data points for correlative analysis or model training, the ASE-based approach can predict splicing-associated genetic variants using RNA-Seq data of a single individual. Thus, it is both cost-effective and computationally inexpensive.

2.3.3 *Trans-acting Regulators of Alternative Splicing*

2.3.3.1 Methods for Identification of Splicing Factors

Recently, an increasing number of RNA-binding proteins (RBPs) have been identified as regulators of splicing [98]. However, the associated splicing factors are not yet known for a large number of SREs identified using the experimental or bioinformatic methods described above. To this end, a modified RNA affinity purification method was used to identify *trans*-factors for known SREs [81, 82, 114]. In addition, in vivo siRNA screens targeting known splicing factors were also used to reveal the *trans*-factor for specific SREs [79, 115–117].

Previous efforts were also dedicated to predict or validate proteins with splicing regulatory activity [98]. For example, a computational pipeline was designed to search for proteins with splicing factor-like properties, which led to the discovery of an SR-related protein with important function in neuronal tissues [118]. Given a pool of RBPs, a previous study screened for splicing-related ones by examining the correlation of their expression with changes in levels of alternative splicing [119]. Combined with motif analysis, the authors successfully identified known and novel splicing factors.

2.3.3.2 Methods for Identifying Binding Motifs of Splicing Factors

Given a splicing factor or RBP, a number of experimental methods were developed to identify their binding motifs globally. These methods can be largely categorized into two classes depending on their in vitro or in vivo nature. The Systematic Evolution of Ligands by EXponential enrichment (SELEX) approach is one of the in vitro methods [120]. SELEX was applied to identify ESEs and other SREs in several studies [121]. Recently, this method was combined with microarray assays to increase the throughput [122]. Another in vitro method called RNAcompete uses in vitro transcribed RNA (structured or unstructured) for pull-down with an RBP of interest, followed by microarray analysis of the bound RNA [123]. Binding motifs of over 200 RBPs were determined by this method [119]. More recently, a new in vitro method called RNA Bind-n-Seq (RBNS) was developed to improve quantification of the sequence and structural specificity of RBPs [124]. Besides canonical motifs, RBNS identified additional near-optimal binding motifs, which were shown to be functional in vivo [124].

To identify global in vivo binding sites of RBPs, the most widely used method is UV cross-linking and immunoprecipitation (CLIP) followed by sequencing (CLIP-Seq) [125]. Variations of this method are also used for different applications, including high-throughput sequencing of RNAs isolated by CLIP (HITS-CLIP) [126], photoactivatable-ribonucleoside-enhanced cross-linking and precipitation (PAR-CLIP) [127], and individual-nucleotide resolution UV cross-linking and immunoprecipitation (iCLIP) [128]. Detailed discussions of these methods are provided by previous reviews [129, 130]. Briefly, CLIP-based methods have relatively high sensitivity and specificity compared to RNA immunoprecipitation alone. However, the cross-linking efficiency is generally limited in regular CLIP, which is improved in PAR-CLIP via the usage of 4-thiouridine, a photo-activated nucleotide. Deletions, substitutions, or insertions usually occur near the cross-linking sites in CLIP-Seq/HITS-CLIP [131], whereas T-to-C substitutions are observed near the cross-linking sites in PAR-CLIP. These mutations can serve as diagnostic features to pinpoint binding sites. Nonetheless, accurate read mapping tolerating such mutations is challenging. Currently, bioinformatic tools are designed to handle read mapping, cluster calling, and motif enrichment. In the future, development of tools that integrate these basic analyses with RNA secondary structure, evolutionary conservation, and in vitro binding data will tremendously facilitate a systematic understanding of protein–RNA interaction.

Notably, the ENCODE Consortium has devoted great efforts to generate CLIP-Seq data of about 200 RBPs. In addition, shRNA knockdown experiments of each RBP are carried out followed by RNA-Seq in cultured cells (K562 and HepG2). These data sets will facilitate identification of splicing regulatory motifs, analysis of splicing factor functions, and generation of global regulatory maps of these RBPs.

2.3.4 Splicing Code

While most existing methods focus mainly on one or a few aspects of splicing regulation, Barash et al. took a step further to assemble a "splicing code" by integrating hundreds of RNA features and the alternative splicing patterns of a wide panel of tissues [109]. This model takes as input exon sequences of interest and their flanking introns, and recursively selects for features and parameters that maximize the "code quality." The code was later improved using Bayesian neural networks on an expanded list of RNA features [132, 133] and applied to predict splicing-altering disease mutations [108]. The above work mainly focused on analysis of alternatively skipped exons. A more recent splicing code was designed to identify RNA sequence features that categorize several major classes of alternative splicing, including exon skipping, alternative 5′ss, and alternative 3′ss exons [134]. This work demonstrated that RNA sequence features (splice sites, conservation levels, and exon/intron architecture) confer strong discriminatory contributions to classify different types of splicing.

Current versions of the splicing code are not able to predict absolute levels of exon inclusion, but rather focus on predictions of relative changes in splicing across tissues or in the presence of genetic mutations. Future development of the splicing code could be empowered by consideration of regulatory networks of multiple splicing factors, epigenetic influence, and kinetic aspects of splicing, some of which are discussed below.

2.3.5 Useful Databases

Over the years, the splicing community has built many databases and Web resources to include data on global profiling of alternative splicing and systematic analysis of splicing regulatory mechanisms. Table 2.2 summarizes some of these resources ranging from catalogs of alternative splicing events to disease-related mutations that affect splicing.

Table 2.2 Databases (DB) of alternative splicing events and software programs (PR) for studies of splicing regulation

Type	Category	Database	URL	Function	Input	Notes
DB	AS events	DBASS	http://www.dbass.org.uk	Catalogs splice site mutations and diseases	Chosen from user drop-down menu	Includes DBASS3 and DBASS5
DB	Data	HGMD	http://www.hgmd.cf.ac.uk/ac/index.php	Provides disease-associated variants	–	Requires license to access the full database
DB	Data	ENCODE	https://www.encodeproject.org	Provides various NGS data sets	Chosen from user drop-down menu	–
DB	Data	RBPDB	http://rbpdb.ccbr.utoronto.ca	Provides RBP PWMs	–	–
DB	Data	CisBP-RNA	http://cisbp-rna.ccbr.utoronto.ca	Provides RBP PWMs	–	–
DB	Data	CLIPdb	http://lulab.life.tsinghua.edu.cn/clipdb	Catalogs references for RBP studies and CLIP data sets	RBP names, cell types, species, or technology types	Also supports database browsing without requiring input
DB	Data	CLIPZ	http://www.clipz.unibas.ch	Provides CLIP data sets with target predictions of RBPs and miRNAs	Chosen from user drop-down menu	Require account sign-up; review [168]
DB	Data	doRINA 2.0	http://dorina.mdc-berlin.de	Provides CLIP data sets with RBP sites annotations	Chosen from user drop-down menu	Review [168]
DB	Data	StarBase V2.0	http://starbase.sysu.edu.cn/index.php	Provides CLIP data sets with annotations	Chosen from user drop-down menu	Review [168]
DB	Tools	OMICtools	http://omictools.com	Databases of genomic, transcriptomic, proteomic, and metabolomic tools	Tool names, analysis type, or Web-browsing	–
DB	Regulation	RegulomeDB	http://regulomedb.org	Displays various regulatory tracks for the input sequences	dbSNP IDs, genomic coordinates in BED files, VCF files, or GFF3 files	Links regulation to GWAS: http://regulomedb.org/GWAS/index.html

(continued)

Table 2.2 (continued)

Type	Category	Database	URL	Function	Input	Notes
DB	Regulation	RegRNA2.0	http://regrna2.mbc.nctu.edu.tw	Displays various regulatory tracks for the input sequences	RNA sequences in FASTA format	–
DB	Regulation	Brain RNA-Seq	http://web.stanford.edu/group/barres_lab/brain_rnaseq.html	Provides FPKM data in various brain cells; compares gene enrichment across available cell types	Gene name, or cell types	Done in mouse; also provides data browsing
DB	Regulation	rSNPBase	http://rsnp.psych.ac.cn	Provides regulatory annotation for SNPs	SNP ID or gene names	–
PR	Regulation	CRYP-SKIP	http://cryp-skip.img.cas.cz	Predicts splice site mutations	Nucleotide sequence in FASTA format	–
PR	Regulation	EX-SKIP	http://ex-skip.img.cas.cz	Compares a SNV's impact on ESE/ESS to induce exon skipping	Two exonic sequences in FASTA format	Maximum 4000nt per submission
PR	Regulation	Human Splicing Finder	http://www.umd.be/HSF	Combines 12 different algorithms to predict mutations' impact on *cis*-regulatory elements	Chosen from user drop-down menu	–
PR	Regulation	MaxEntScan	http://genes.mit.edu/burgelab/maxent/Xmaxentscan_scoreseq.html; http://genes.mit.edu/burgelab/maxent/Xmaxentscan_scoreseq_acc.html	Predicts splice site strength	Splice site sequences in FASTA format	Web-based or command line-based analyses available

(continued)

Table 2.2 (continued)

Type	Category	Database	URL	Function	Input	Notes
PR	Regulation	WASP	http://genes.toronto.edu/wasp	Predicts AS exons and the potential regulation codes	RNA sequence in FASTA format or genomic coordinates in BED format	Maximum 10 input exons per query
PR	Regulation	AVISPA	http://avispa.biociphers.org	Predicts AS exons and the potential regulation codes	FASTA or BED files containing a single putative AS exon or cassette exon triplet	–
PR	Regulation	SPANR	http://tools.genes.toronto.edu	Predicts SNVs effects on cassette exons	Maximum 40 SNVs per file in tab-delimited VCF format	This tool was designed for detecting exon-skipping events, so it may or may not work for other AS types

2.4 Ongoing Questions

2.4.1 *Gene Expression Kinetics and Co-Transcriptional Splicing*

With the advent of RNA-Seq and related methodologies described previously in this chapter, it is now possible to study kinetics of gene expression and splicing on the global scale. It was recently shown that several steps in RNA processing often, but not always, occur co-transcriptionally, including capping, splicing, and polyadenylation, allowing for efficient and accurate pre-mRNA maturation (reviewed in [121, 135, 136]). In particular, co-transcriptional splicing depends on the rate of RNA Pol II elongation with the idea that slower elongation allows more time for splicing to complete. Pol II elongation can be affected by nucleosome positioning, DNA methylation, histone modifications, and chromatin remodeling [137, 138] (Fig. 2.3). Additionally, the C-terminal domain of RNA Pol II can be post-translationally and reversibly modified to guide interactions with different proteins involved in RNA processing. Thus, chromatin modifications, transcription, and splicing are all interconnected processes [136, 137].

To study dynamic regulation of gene expression and/or co-transcriptional splicing, nascent RNA must be captured. Modified RNA-Seq methods such as genomic run-on sequencing (GRO-Seq) or sequencing of 4-thiouridine-labeled RNA may be analyzed in conjunction with RNA-Seq [139–141]. Additionally, cell fractionation and selection of non-polyadenylated RNA in the chromatin fraction may be used. Recently, a native elongating transcript sequencing (NET-Seq) approach was used by two groups to identify spliceosome-mediated cleavage, Pol II dynamics related to splicing, and antisense transcription [142, 143]. Figure 2.3 illustrates co-transcriptional splicing and other events described below including RNA editing, mirtron biogenesis, and circRNA biogenesis.

2.4.2 *RNA Modifications*

RNA modifications such as methylation (primarily N^6-methyladenosine, or m6A) and RNA editing were not extensively studied until recently. m6A, originally identified in tRNAs, rRNAs, and snoRNAs, was recently shown to be widespread in mRNAs with potential impact on splicing, mRNA degradation, and RNA secondary structures [144, 145]. The most prevalent form of RNA editing is the conversion of adenosine to inosine (A-to-I) via deamination, typically in double-stranded RNA (dsRNA) regions by adenosine deaminases acting on RNA (ADARs) (Fig. 2.3). In order for editing to affect splicing, it is expected to occur before splicing is completed. Indeed, several lines of evidence suggest editing

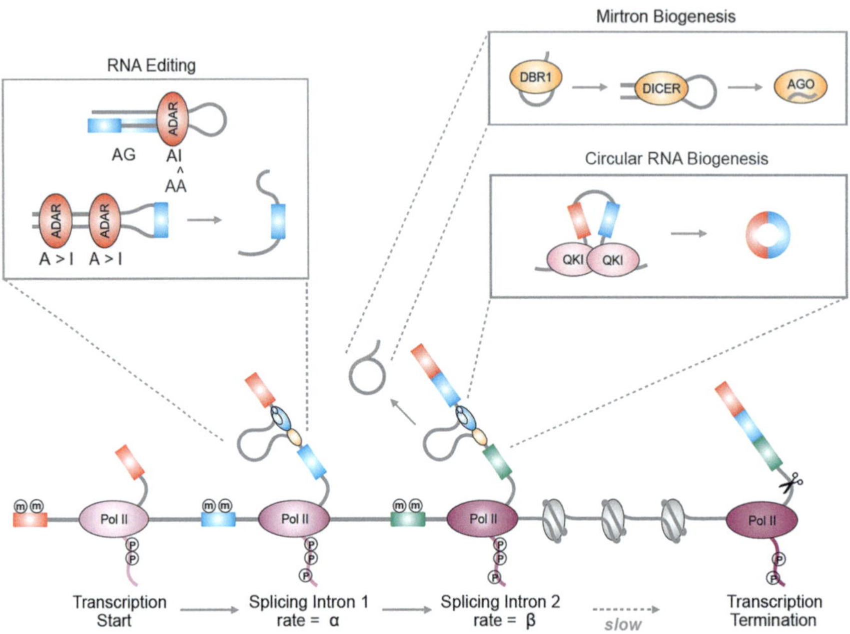

Fig. 2.3 Co-transcriptional splicing and related RNA products. Co-transcriptional splicing of two introns with splicing rates α and β is shown. The following epigenetic factors are illustrated: DNA methylation (m) enrichment in exons [138], dynamic phosphorylation (P) of the C-terminal domain of RNA Pol II [137], and nucleosomes slowing down Pol II transcription. Splicing coupled with RNA editing and the biogenesis of mirtrons and circRNAs are shown in the insets. RNA editing can generate new splice sites (e.g., changing A to I may create a new AG 3′ss, RNA editing inset, top [146]) and prevent circRNA biogenesis (RNA editing inset, bottom [151]). Mirtrons are derived from lariats that are debranched by DBR1 and processed by DICER (mirtron biogenesis inset [147]). QKI regulates the production of a subset of circRNAs (circRNA Biogenesis inset [155])

precedes splicing (reviewed in [146]), although exceptions do exist. These findings are only the beginning of a new era of functional and mechanistic studies of RNA modifications.

2.4.3 Splicing Generates Other RNA Species

Although introns are typically degraded after removal, certain introns can also be further processed to generate other RNA species. For example, biogenesis pathways of snoRNAs, mirtrons, and simtrons rely on intron splicing (reviewed in [147]). Whereas canonical miRNA biogenesis depends on the microprocessor (DGCR8 and DROSHA), mirtrons depend on lariat debranching (Fig. 2.3) and simtrons depend on U1 snRNP. Another RNA species underappreciated until recently are

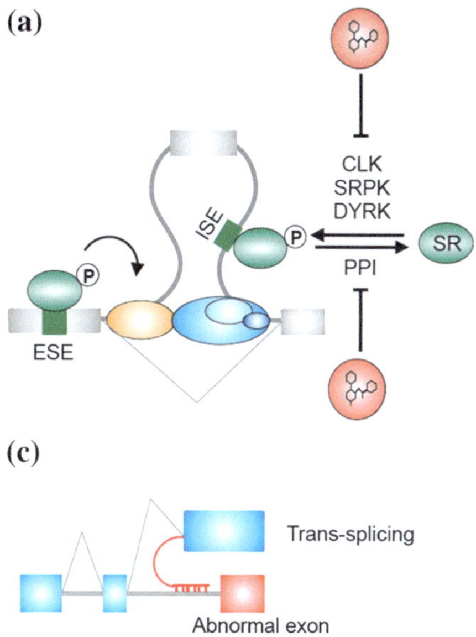

Fig. 2.4 Therapeutic approaches to modulate splicing. **a** Small molecule therapy. Phosphorylation or dephosphorylation of SR proteins are regulated by CDC2-like kinase (CLK), dual-specificity tyrosine-(Y)-phosphorylation-regulated kinase (DYRK), SR protein kinase (SRPK), and protein phosphatase-1 (PP1). Inhibitors of these kinases and phosphatase affect the associated splicing events. **b** Antisense oligonucleotides therapy. SR protein or other splicing factor binding sites can be blocked by ASO to achieve specific alternation of splicing. **c** Trans-splicing therapy. ASO linked to a restoring normal exon can rescue an abnormal splicing event that may result due to multiple mutations

circular RNAs (circRNAs) (reviewed in [148, 149]). It was shown that biogenesis of certain circRNAs depends on intronic sequence content [150–152], which may compete with pre-mRNA splicing [153]. Additionally, circRNAs can contain both exons and introns, and two of these were shown to regulate gene expression [154]. The splicing factor QKI was shown to regulate production of many circRNAs (Fig. 2.3) [155]. The biogenesis and functions of circRNAs are currently under active investigation.

2.4.4 Global Misregulation of Splicing in Disease

Since splicing is required for RNA maturation, misregulation of splicing may lead to disease states [11]. In addition to well-known splicing diseases, such as myotonic dystrophy [156], there are several examples of point mutations in specific genes that cause splicing misregulation (reviewed in [121, 157]). Furthermore, global splicing misregulation also characterizes some diseases such as cancer. The Cancer Genome Atlas (TCGA, www.cancergenome.nih.gov) provides a wealth of genomic data from cancer patients and controls, allowing for the study of global splicing alterations within and across cancer types [158, 159]. Splicing abnormalities were also shown in autistic brains [160]. Although splicing alterations in cancer are well established, it is difficult to identify the mechanistic cause and functional significance of these events, especially considering that up to hundreds of RBPs may be involved in the regulation of thousands of alternative splicing events in both normal and disease states [121, 157]. In the future, an understanding of the causes and functional consequences may lead to splicing-targeted therapeutics.

2.5 Splicing as a Therapeutic Target

Given the critical roles of splicing misregulation in disease, a number of strategies are under development to therapeutically correct aberrant splicing events. First, small molecules can be used to directly modulate the activity of splicing factors [161]. The advantage of this method is the ease of delivery and the potential for individual-specific dosage control. As examples, small molecule inhibitors were examined that target SR protein kinases (SRPKs), CDC2-like kinases (CLKs), or protein phosphatase-1 (PP1), which can then modulate phosphorylation of SR proteins (Fig. 2.4). However, such inhibitors often have off-target effects and affect splicing of many genes.

A more targeted approach involves usage of antisense oligonucleotides (ASO), reverse complementary sequences that bind to target mRNA sequences. Because ASOs are sequence-specific, they can block binding of splicing factors at specific loci and modulate alternative splicing. For example, aberrant splicing events caused by an intronic mutation in the human β-globin gene were corrected by ASO treatment in a β-thalassemia mouse model [162]. In addition, clinical trials are underway for ASO-based therapy of Duchenne muscular dystrophy and spinal muscular atrophy [163]. Although ASO therapy overcomes the nonspecificity issue of small molecules, their delivery is relatively difficult. Another method, trans-splicing, is an effective strategy for repairing multiple mutations in exons or transcripts. Also referred to as Spliceosomal-mediated RNA trans-splicing (SMaRT) [164], this method can replace the entire mRNA sequence 5′ or 3′ of a target splice site by trans-splicing between an ASO and the endogenous RNA [165]. This approach was proposed as a therapy for β-thalassemia to replace the first exon of the β-globin gene resulting from aberrant splicing [166]. However, the delivery of trans-splicing therapy is also challenging, as it necessitates incorporation of DNA vectors to cells (10).

2.6 Conclusions

In recent years, technological advances brought a fundamental shift in our approaches to splicing-related questions. Global analyses that combine high-throughput experimental assays and bioinformatic methods are becoming indispensible. As a result, numerous novel insights have been revealed regarding the landscape of alternative splicing and the regulatory mechanisms of splicing in various cell types. These global discoveries constitute a foundation for further mechanistic and functional studies in model systems and translational research. However, there still exist many challenges in handling high-throughput experiments and data analysis. We expect that these challenges will be addressed via methodology development and standardization, which will further catalyze exciting discoveries in splicing research.

References

1. Berget SM, Moore C, Sharp PA. Spliced segments at the 5′ terminus of adenovirus 2 late mRNA. Proc Natl Acad Sci USA. 1977;74(8):3171–5.
2. Chow LT, Gelinas RE, Broker TR, Roberts RJ. An amazing sequence arrangement at the 5′ ends of adenovirus 2 messenger RNA. Cell. 1977;12(1):1–8.
3. Wahl MC, Will CL, Luhrmann R. The spliceosome: design principles of a dynamic RNP machine. Cell. 2009;136(4):701–18.
4. Wang ET, Sandberg R, Luo S, Khrebtukova I, Zhang L, Mayr C, Kingsmore SF, Schroth GP, Burge CB. Alternative isoform regulation in human tissue transcriptomes. Nature. 2008;456(7221):470–6.
5. Pan Q, Shai O, Lee LJ, Frey BJ, Blencowe BJ. Deep surveying of alternative splicing complexity in the human transcriptome by high-throughput sequencing. Nat Genet. 2008;40(12):1413–5.
6. Nilsen TW, Graveley BR. Expansion of the eukaryotic proteome by alternative splicing. Nature. 2010;463:457–63.
7. Schmucker D, Clemens JC, Shu H, Worby CA, Xiao J, Muda M, Dixon JE, Zipursky SL. Drosophila Dscam is an axon guidance receptor exhibiting extraordinary molecular diversity. Cell. 2000;101(6):671–84.
8. Matlin AJ, Clark F, Smith CW. Understanding alternative splicing: towards a cellular code. Nat Rev Mol Cell Biol. 2005;6(5):386–98.
9. Kalsotra A, Cooper TA. Functional consequences of developmentally regulated alternative splicing. Nat Rev Genet. 2011;12(10):715–29.
10. Wang GS, Cooper TA. Splicing in disease: disruption of the splicing code and the decoding machinery. Nat Rev Genet. 2007;8(10):749–61.
11. Cooper TA, Wan L, Dreyfuss G. RNA and disease. Cell. 2009;136(4):777–93.
12. Poulos MG, Batra R, Charizanis K, Swanson MS. Developments in RNA splicing and disease. Cold Spring Harb Perspect Biol. 2011;3(1):a000778.
13. Black DL. Mechanisms of alternative pre-messenger RNA splicing. Annu Rev Biochem. 2003;72:291–336.
14. Buratti E, Baralle FE. Influence of RNA secondary structure on the pre-mRNA splicing process. Mol Cell Biol. 2004;24(24):10505–14.

15. Lee C, Roy M. Analysis of alternative splicing with microarrays: successes and challenges. Genome Biol. 2004;5(7):231.
16. Cuperlovic-Culf M, Belacel N, Culf AS, Ouellette RJ. Microarray analysis of alternative splicing. OMICS. 2006;10(3):344–57.
17. Blencowe BJ. Alternative splicing: new insights from global analyses. Cell. 2006;126 (1):37–47.
18. Hu GK, Madore SJ, Moldover B, Jatkoe T, Balaban D, Thomas J, Wang Y. Predicting splice variant from DNA chip expression data. Genome Res. 2001;11(7):1237–45.
19. Johnson JM, Castle J, Garrett-Engele P, Kan Z, Loerch PM, Armour CD, Santos R, Schadt EE, Stoughton R, Shoemaker DD. Genome-wide survey of human alternative pre-mRNA splicing with exon junction microarrays. Science. 2003;302(5653):2141–4.
20. Pan Q, Shai O, Misquitta C, Zhang W, Saltzman AL, Mohammad N, Babak T, Siu H, Hughes TR, Morris QD, et al. Revealing global regulatory features of mammalian alternative splicing using a quantitative microarray platform. Mol Cell. 2004;16(6):929–41.
21. Wang Z, Gerstein M, Snyder M. RNA-Seq: a revolutionary tool for transcriptomics. Nat Rev Genet. 2009;10(1):57–63.
22. Sultan M, Schulz MH, Richard H, Magen A, Klingenhoff A, Scherf M, Seifert M, Borodina T, Soldatov A, Parkhomchuk D, et al. A global view of gene activity and alternative splicing by deep sequencing of the human transcriptome. Science. 2008;321 (5891):956–60.
23. Lee JH, Gao C, Peng G, Greer C, Ren S, Wang Y, Xiao X. Analysis of transcriptome complexity through RNA sequencing in normal and failing murine hearts. Circ Res. 2011;109(12):1332–41.
24. Mortazavi A, Williams BA, McCue K, Schaeffer L, Wold B. Mapping and quantifying mammalian transcriptomes by RNA-Seq. Nat Methods. 2008;5(7):621–8.
25. Li G, Bahn JH, Lee JH, Peng G, Chen Z, Nelson SF, Xiao X. Identification of allele-specific alternative mRNA processing via transcriptome sequencing. Nucleic Acids Res. 2012;40 (13):e104.
26. Majewski J, Pastinen T. The study of eQTL variations by RNA-seq: from SNPs to phenotypes. Trends Genet: [TIG]. 2011;27(2):72–9.
27. Wulff BE, Sakurai M, Nishikura K. Elucidating the inosinome: global approaches to adenosine-to-inosine RNA editing. Nat Rev Genet. 2011;12(2):81–5.
28. Bahn JH, Lee JH, Li G, Greer C, Peng G, Xiao X. Accurate identification of A-to-I RNA editing in human by transcriptome sequencing. Genome Res. 2012;22(1):142–50.
29. Lee JH, Ang JK, Xiao X. Analysis and design of RNA sequencing experiments for identifying RNA editing and other single-nucleotide variants. RNA. 2013;19(6):725–32.
30. Zhang Q, Xiao X. Genome sequence-independent identification of RNA editing sites. Nat Methods. 2015;12(4):347–50.
31. Kratz A, Carninci P. The devil in the details of RNA-seq. Nat Biotechnol. 2014;32(9):882–4.
32. van Dijk EL, Jaszczyszyn Y, Thermes C. Library preparation methods for next-generation sequencing: tone down the bias. Exp Cell Res. 2014;322(1):12–20.
33. Head SR, Komori HK, LaMere SA, Whisenant T, Van Nieuwerburgh F, Salomon DR, Ordoukhanian P. Library construction for next-generation sequencing: overviews and challenges. BioTech 2014;56(supp 2):61–4, 66, 68, passim.
34. Leek JT, Scharpf RB, Bravo HC, Simcha D, Langmead B, Johnson WE, Geman D, Baggerly K, Irizarry RA. Tackling the widespread and critical impact of batch effects in high-throughput data. Nat Rev Genet. 2010;11(10):733–9.
35. Liu Y, Ferguson JF, Xue C, Silverman IM, Gregory B, Reilly MP, Li M. Evaluating the impact of sequencing depth on transcriptome profiling in human adipose. PLoS ONE. 2013; 8(6):e66883.
36. Katz Y, Wang ET, Airoldi EM, Burge CB. Analysis and design of RNA sequencing experiments for identifying isoform regulation. Nat Methods. 2010;7(12):1009–15.

37. Li H, Qiu J, Fu XD. RASL-seq for massively parallel and quantitative analysis of gene expression. In: Frederick M Ausubel et al. (Ed.) Current protocols in molecular biology, 2012; Chap. 4:Unit 4 13, pp 11–9.
38. Li H, Zhou H, Wang D, Qiu J, Zhou Y, Li X, Rosenfeld MG, Ding S, Fu XD. Versatile pathway-centric approach based on high-throughput sequencing to anticancer drug discovery. Proc Natl Acad Sci USA. 2012;109(12):4609–14.
39. Larman HB, Scott ER, Wogan M, Oliveira G, Torkamani A, Schultz PG. Sensitive, multiplex and direct quantification of RNA sequences using a modified RASL assay. Nucleic Acids Res. 2014;42(14):9146–57.
40. Garber M, Grabherr MG, Guttman M, Trapnell C. Computational methods for transcriptome annotation and quantification using RNA-seq. Nat Methods. 2011;8(6):469–77.
41. Steijger T, Abril JF, Engstrom PG, Kokocinski F, Hubbard TJ, Guigo R, Harrow J, Bertone P, Consortium R. Assessment of transcript reconstruction methods for RNA-seq. Nat Methods. 2013;10(12):1177–84.
42. Roy CK, Olson S, Graveley BR, Zamore PD, Moore MJ. Assessing long-distance RNA sequence connectivity via RNA-templated DNA–DNA ligation. eLife 2015;4.
43. Zhang F, Wang M, Michael T, Drabier R. Novel alternative splicing isoform biomarkers identification from high-throughput plasma proteomics profiling of breast cancer. BMC Syst Biol. 2013;7(Suppl 5):S8.
44. Chen L. Statistical and computational methods for high-throughput sequencing data analysis of alternative splicing. Stat Biosci. 2013;5(1):138–55.
45. Hooper JE. A survey of software for genome-wide discovery of differential splicing in RNA-Seq data. Hum Genomics. 2014;8:3.
46. Patro R, Mount SM, Kingsford C. Sailfish enables alignment-free isoform quantification from RNA-seq reads using lightweight algorithms. Nat Biotechnol. 2014;32(5):462–4.
47. Zhang Z, Wang W. RNA-Skim: a rapid method for RNA-Seq quantification at transcript level. Bioinformatics. 2014;30(12):i283–92.
48. Aschoff M, Hotz-Wagenblatt A, Glatting KH, Fischer M, Eils R, Konig R. SplicingCompass: differential splicing detection using RNA-seq data. Bioinformatics. 2013;29(9):1141–8.
49. Shen S, Park JW, Lu Z-X, Lin L, Henry MD, Wu YN, Zhou Q, Xing Y. rMATS: robust and flexible detection of differential alternative splicing from replicate RNA-Seq data. Proc Natl Acad Sci USA. 2014;111(51):E5593–601.
50. Kahles A, Ong CS, Ratsch G. SplAdder: identification, quantification and testing of alternative splicing events from RNA-Seq data. Biorxiv. 2015:017095.
51. Hu Y, Huang Y, Du Y, Orellana CF, Singh D, Johnson AR, Monroy A, Kuan PF, Hammond SM, Makowski L, et al. DiffSplice: the genome-wide detection of differential splicing events with RNA-seq. Nucleic Acids Res. 2013;41(2):e39.
52. Guttman M, Garber M, Levin JZ, Donaghey J, Robinson J, Adiconis X, Fan L, Koziol MJ, Gnirke A, Nusbaum C, et al. Ab initio reconstruction of cell type-specific transcriptomes in mouse reveals the conserved multi-exonic structure of lincRNAs. Nat Biotechnol. 2010; 28(5):503–10.
53. Trapnell C, Williams BA, Pertea G, Mortazavi A, Kwan G, van Baren MJ, Salzberg SL, Wold BJ, Pachter L. Transcript assembly and quantification by RNA-Seq reveals unannotated transcripts and isoform switching during cell differentiation. Nat Biotechnol. 2010;28(5):511–5.
54. Grabherr MG, Haas BJ, Yassour M, Levin JZ, Thompson DA, Amit I, Adiconis X, Fan L, Raychowdhury R, Zeng Q, et al. Full-length transcriptome assembly from RNA-Seq data without a reference genome. Nat Biotechnol. 2011;29(7):644–52.
55. Li B, Dewey CN. RSEM: accurate transcript quantification from RNA-Seq data with or without a reference genome. BMC Bioinform. 2011;12:323.
56. Suo C, Calza S, Salim A, Pawitan Y. Joint estimation of isoform expression and isoform-specific read distribution using multisample RNA-Seq data. Bioinformatics. 2014; 30(4):506–13.

57. Hu Y, Liu Y, Mao X, Jia C, Ferguson JF, Xue C, Reilly MP, Li H, Li M. PennSeq: accurate isoform-specific gene expression quantification in RNA-Seq by modeling non-uniform read distribution. Nucleic Acids Res. 2014;42(3):e20.
58. Wang Z, Lo HS, Yang H, Gere S, Hu Y, Buetow KH, Lee MP. Computational analysis and experimental validation of tumor-associated alternative RNA splicing in human cancer. Cancer Res. 2003;63(3):655–7.
59. Wang Z, Rolish ME, Yeo G, Tung V, Mawson M, Burge CB. Systematic identification and analysis of exonic splicing silencers. Cell. 2004;119(6):831–45.
60. Xiao X, Wang Z, Jang M, Burge CB. Coevolutionary networks of splicing cis-regulatory elements. Proc Natl Acad Sci USA. 2007;104(47):18583–8.
61. Xiao X, Wang Z, Jang M, Nutiu R, Wang ET, Burge CB. Splice site strength-dependent activity and genetic buffering by poly-G runs. Nat Struct Mol Biol. 2009;16(10):1094–100.
62. Mark D, Haeberle S, Roth G, von Stetten F, Zengerle R. Microfluidic lab-on-a-chip platforms: requirements, characteristics and applications. Chem Soc Rev. 2010;39(3):1153–82.
63. Arias MA, Lubkin A, Chasin LA. Splicing of designer exons informs a biophysical model for exon definition. RNA. 2015;21(2):213–29.
64. Jian X, Boerwinkle E, Liu X. In silico tools for splicing defect prediction: a survey from the viewpoint of end users. Genet Med: Off J Am Coll Med Genet. 2014;16(7):497–503.
65. Desmet FO, Beroud C. Bioinformatics and mutations leading to exon skipping. Methods Mol Biol. 2012;867:17–35.
66. Desmet FO, Hamroun D, Lalande M, Collod-Beroud G, Claustres M, Beroud C. Human splicing finder: an online bioinformatics tool to predict splicing signals. Nucleic Acids Res. 2009;37(9):e67.
67. Reese MG, Eeckman FH, Kulp D, Haussler D. Improved splice site detection in genie. J Comput Biol: J Comput Mol Cell Biol. 1997;4(3):311–23.
68. Yeo G, Burge CB. Maximum entropy modeling of short sequence motifs with applications to RNA splicing signals. J Comput Biol: J Comput Mol Cell Biol. 2004;11(2–3):377–94.
69. Gooding C, Clark F, Wollerton MC, Grellscheid SN, Groom H, Smith CW. A class of human exons with predicted distant branch points revealed by analysis of AG dinucleotide exclusion zones. Genome Biol. 2006;7(1):R1.
70. Gao K, Masuda A, Matsuura T, Ohno K. Human branch point consensus sequence is yUnAy. Nucleic Acids Res. 2008;36(7):2257–67.
71. Plass M, Agirre E, Reyes D, Camara F, Eyras E. Co-evolution of the branch site and SR proteins in eukaryotes. Trends Genet: TIG. 2008;24(12):590–4.
72. Schwartz SH, Silva J, Burstein D, Pupko T, Eyras E, Ast G. Large-scale comparative analysis of splicing signals and their corresponding splicing factors in eukaryotes. Genome Res. 2008;18(1):88–103.
73. Corvelo A, Hallegger M, Smith CW, Eyras E. Genome-wide association between branch point properties and alternative splicing. PLoS Comput Biol. 2010;6(11):e1001016.
74. Pastuszak AW, Joachimiak MP, Blanchette M, Rio DC, Brenner SE, Frankel AD. An SF1 affinity model to identify branch point sequences in human introns. Nucleic Acids Res. 2011;39(6):2344–56.
75. Bitton DA, Rallis C, Jeffares DC, Smith GC, Chen YY, Codlin S, Marguerat S, Bahler J. LaSSO, a strategy for genome-wide mapping of intronic lariats and branch points using RNA-seq. Genome Res. 2014;24(7):1169–79.
76. Taggart AJ, DeSimone AM, Shih JS, Filloux ME, Fairbrother WG. Large-scale mapping of branchpoints in human pre-mRNA transcripts in vivo. Nat Struct Mol Biol. 2012; 19(7):719–21.
77. Mercer TR, Gerhardt DJ, Dinger ME, Crawford J, Trapnell C, Jeddeloh JA, Mattick JS, Rinn JL. Targeted RNA sequencing reveals the deep complexity of the human transcriptome. Nat Biotechnol. 2012;30(1):99–104.

78. Mercer TR, Clark MB, Andersen SB, Brunck ME, Haerty W, Crawford J, Taft RJ, Nielsen LK, Dinger ME, Mattick JS. Genome-wide discovery of human splicing branchpoints. Genome Res. 2015;25(2):290–303.
79. Culler SJ, Hoff KG, Voelker RB, Berglund JA, Smolke CD. Functional selection and systematic analysis of intronic splicing elements identify active sequence motifs and associated splicing factors. Nucleic Acids Res. 2010;38(15):5152–65.
80. Ke S, Shang S, Kalachikov SM, Morozova I, Yu L, Russo JJ, Ju J, Chasin LA. Quantitative evaluation of all hexamers as exonic splicing elements. Genome Res. 2011;21(8):1360–74.
81. Wang Y, Ma M, Xiao X, Wang Z. Intronic splicing enhancers, cognate splicing factors and context-dependent regulation rules. Nat Struct Mol Biol. 2012;19(10):1044–52.
82. Wang Y, Xiao X, Zhang J, Choudhury R, Robertson A, Li K, Ma M, Burge CB, Wang Z. A complex network of factors with overlapping affinities represses splicing through intronic elements. Nat Struct Mol Biol. 2013;20(1):36–45.
83. Fairbrother WG, Yeh RF, Sharp PA, Burge CB. Predictive identification of exonic splicing enhancers in human genes. Science. 2002;297(5583):1007–13.
84. Zhang XH, Chasin LA. Computational definition of sequence motifs governing constitutive exon splicing. Genes Dev. 2004;18(11):1241–50.
85. Yeo G, Hoon S, Venkatesh B, Burge CB. Variation in sequence and organization of splicing regulatory elements in vertebrate genes. Proc Natl Acad Sci USA. 2004;101(44):15700–5.
86. Venkatesh B, Yap WH. Comparative genomics using fugu: a tool for the identification of conserved vertebrate cis-regulatory elements. Bioessays: News Rev Mol, Cell Dev Biol. 2005;27(1):100–7.
87. Cartegni L, Wang J, Zhu Z, Zhang MQ, Krainer AR. ESEfinder: a web resource to identify exonic splicing enhancers. Nucleic Acids Res. 2003;31(13):3568–71.
88. Zhang XH, Heller KA, Hefter I, Leslie CS, Chasin LA. Sequence information for the splicing of human pre-mRNA identified by support vector machine classification. Genome Res. 2003;13(12):2637–50.
89. Stadler MB, Shomron N, Yeo GW, Schneider A, Xiao X, Burge CB. Inference of splicing regulatory activities by sequence neighborhood analysis. PLoS Genet. 2006;2(11):e191.
90. Zhang J, Kuo CC, Chen L. VERSE: a varying effect regression for splicing elements discovery. J Comput Biol: J Comput Mol Cell Biol. 2012;19(6):855–65.
91. Badr E, Heath LS. Identifying splicing regulatory elements with de Bruijn graphs. J Comput Biol: J Comput Mol Cell Biol. 2014;21(12):880–97.
92. Friedman BA, Stadler MB, Shomron N, Ding Y, Burge CB. Ab initio identification of functionally interacting pairs of cis-regulatory elements. Genome Res. 2008;18(10):1643–51.
93. Yu Y, Maroney PA, Denker JA, Zhang XH, Dybkov O, Luhrmann R, Jankowsky E, Chasin LA, Nilsen TW. Dynamic regulation of alternative splicing by silencers that modulate 5′ splice site competition. Cell. 2008;135(7):1224–36.
94. Ke S, Chasin LA. Intronic motif pairs cooperate across exons to promote pre-mRNA splicing. Genome Biol. 2010;11(8):R84.
95. Weyn-Vanhentenryck SM, Mele A, Yan Q, Sun S, Farny N, Zhang Z, Xue C, Herre M, Silver PA, Zhang MQ, et al. HITS-CLIP and integrative modeling define the Rbfox splicing-regulatory network linked to brain development and autism. Cell Rep. 2014;6(6):1139–52.
96. Zhang C, Frias MA, Mele A, Ruggiu M, Eom T, Marney CB, Wang H, Licatalosi DD, Fak JJ, Darnell RB. Integrative modeling defines the Nova splicing-regulatory network and its combinatorial controls. Science. 2010;329(5990):439–43.
97. Han A, Stoilov P, Linares AJ, Zhou Y, Fu XD, Black DL. De novo prediction of PTBP1 binding and splicing targets reveals unexpected features of its RNA recognition and function. PLoS Comput Biol. 2014;10(1):e1003442.
98. Fu XD, Ares M Jr. Context-dependent control of alternative splicing by RNA-binding proteins. Nat Rev Genet. 2014;15(10):689–701.

99. Krawczak M, Reiss J, Cooper DN. The mutational spectrum of single base-pair substitutions in mRNA splice junctions of human genes: causes and consequences. Hum Genet. 1992;90(1–2):41–54.
100. Ars E, Kruyer H, Gaona A, Serra E, Lazaro C, Estivill X. Prenatal diagnosis of sporadic neurofibromatosis type 1 (NF1) by RNA and DNA analysis of a splicing mutation. Prenat Diagn. 1999;19(8):739–42.
101. Teraoka SN, Telatar M, Becker-Catania S, Liang T, Onengut S, Tolun A, Chessa L, Sanal O, Bernatowska E, Gatti RA, et al. Splicing defects in the ataxia-telangiectasia gene, ATM: underlying mutations and consequences. Am J Hum Genet. 1999;64(6):1617–31.
102. Lopez-Bigas N, Audit B, Ouzounis C, Parra G, Guigo R. Are splicing mutations the most frequent cause of hereditary disease? FEBS Lett. 2005;579(9):1900–3.
103. Kwan T, Benovoy D, Dias C, Gurd S, Provencher C, Beaulieu P, Hudson TJ, Sladek R, Majewski J. Genome-wide analysis of transcript isoform variation in humans. Nat Genet. 2008;40(2):225–31.
104. Zhao K, Lu ZX, Park JW, Zhou Q, Xing Y. GLiMMPS: robust statistical model for regulatory variation of alternative splicing using RNA-seq data. Genome Biol. 2013;14(7):R74.
105. Monlong J, Calvo M, Ferreira PG, Guigo R. Identification of genetic variants associated with alternative splicing using sQTLseekeR. Nature Commun. 2014;5:4698.
106. Mort M, Sterne-Weiler T, Li B, Ball EV, Cooper DN, Radivojac P, Sanford JR, Mooney SD. MutPred Splice: machine learning-based prediction of exonic variants that disrupt splicing. Genome Biol. 2014;15(1):R19.
107. Sterne-Weiler T, Howard J, Mort M, Cooper DN, Sanford JR. Loss of exon identity is a common mechanism of human inherited disease. Genome Res. 2011;21(10):1563–71.
108. Xiong HY, Alipanahi B, Lee LJ, Bretschneider H, Merico D, Yuen RK, Hua Y, Gueroussov S, Najafabadi HS, Hughes TR, et al. RNA splicing. The human splicing code reveals new insights into the genetic determinants of disease. Science. 2015;347(6218):1254806.
109. Barash Y, Calarco JA, Gao W, Pan Q, Wang X, Shai O, Blencowe BJ, Frey BJ. Deciphering the splicing code. Nature. 2010;465(7294):53–9.
110. Barash Y, Vaquero-Garcia J, Gonzalez-Vallinas J, Xiong HY, Gao W, Lee LJ, Frey BJ. AVISPA: a web tool for the prediction and analysis of alternative splicing. Genome Biol. 2013;14(10):R114.
111. Pastinen T. Genome-wide allele-specific analysis: insights into regulatory variation. Nat Rev Genet. 2010;11(8):533–8.
112. Degner JF, Marioni JC, Pai AA, Pickrell JK, Nkadori E, Gilad Y, Pritchard JK. Effect of read-mapping biases on detecting allele-specific expression from RNA-sequencing data. Bioinformatics. 2009;25(24):3207–12.
113. Heap GA, Yang JH, Downes K, Healy BC, Hunt KA, Bockett N, Franke L, Dubois PC, Mein CA, Dobson RJ, et al. Genome-wide analysis of allelic expression imbalance in human primary cells by high-throughput transcriptome resequencing. Hum Mol Genet. 2010;19(1):122–34.
114. Wang Y, Wang Z. Systematical identification of splicing regulatory cis-elements and cognate trans-factors. Methods. 2014;65(3):350–8.
115. Izquierdo JM, Majos N, Bonnal S, Martinez C, Castelo R, Guigo R, Bilbao D, Valcarcel J. Regulation of Fas alternative splicing by antagonistic effects of TIA-1 and PTB on exon definition. Mol Cell. 2005;19(4):475–84.
116. Underwood JG, Boutz PL, Dougherty JD, Stoilov P, Black DL. Homologues of the caenorhabditis elegans Fox-1 protein are neuronal splicing regulators in mammals. Mol Cell Biol. 2005;25(22):10005–16.
117. Huelga SC, Vu AQ, Arnold JD, Liang TY, Liu PP, Yan BY, Donohue JP, Shiue L, Hoon S, Brenner S, et al. Integrative genome-wide analysis reveals cooperative regulation of alternative splicing by hnRNP proteins. Cell Rep. 2012;1(2):167–78.

118. Calarco JA, Superina S, O'Hanlon D, Gabut M, Raj B, Pan Q, Skalska U, Clarke L, Gelinas D, van der Kooy D, et al. Regulation of vertebrate nervous system alternative splicing and development by an SR-related protein. Cell. 2009;138(5):898–910.
119. Ray D, Kazan H, Cook KB, Weirauch MT, Najafabadi HS, Li X, Gueroussov S, Albu M, Zheng H, Yang A, et al. A compendium of RNA-binding motifs for decoding gene regulation. Nature. 2013;499(7457):172–7.
120. Tuerk C, Gold L. Systematic evolution of ligands by exponential enrichment: RNA ligands to bacteriophage T4 DNA polymerase. Science. 1990;249(4968):505–10.
121. Lee Y, Rio DC. Mechanisms and regulation of alternative pre-mRNA splicing. Annu Rev Biochem. 2015;84:291–323.
122. Reid DC, Chang BL, Gunderson SI, Alpert L, Thompson WA, Fairbrother WG. Next-generation SELEX identifies sequence and structural determinants of splicing factor binding in human pre-mRNA sequence. RNA. 2009;15(12):2385–97.
123. Ray D, Kazan H, Chan ET, Pena Castillo L, Chaudhry S, Talukder S, Blencowe BJ, Morris Q, Hughes TR. Rapid and systematic analysis of the RNA recognition specificities of RNA-binding proteins. Nat Biotechnol 2009;27(7):667–70.
124. Lambert N, Robertson A, Jangi M, McGeary S, Sharp PA, Burge CB. RNA bind-n-seq: quantitative assessment of the sequence and structural binding specificity of RNA binding proteins. Mol Cell. 2014;54(5):887–900.
125. Ule J, Jensen K, Mele A, Darnell RB. CLIP: a method for identifying protein-RNA interaction sites in living cells. Methods. 2005;37(4):376–86.
126. Licatalosi DD, Mele A, Fak JJ, Ule J, Kayikci M, Chi SW, Clark TA, Schweitzer AC, Blume JE, Wang X, et al. HITS-CLIP yields genome-wide insights into brain alternative RNA processing. Nature. 2008;456(7221):464–9.
127. Hafner M, Landthaler M, Burger L, Khorshid M, Hausser J, Berninger P, Rothballer A, Ascano M Jr, Jungkamp AC, Munschauer M, et al. Transcriptome-wide identification of RNA-binding protein and microRNA target sites by PAR-CLIP. Cell. 2010;141(1):129–41.
128. Konig J, Zarnack K, Rot G, Curk T, Kayikci M, Zupan B, Turner DJ, Luscombe NM, Ule J. iCLIP reveals the function of hnRNP particles in splicing at individual nucleotide resolution. Nat Struct Mol Biol. 2010;17(7):909–15.
129. McHugh CA, Russell P, Guttman M. Methods for comprehensive experimental identification of RNA-protein interactions. Genome Biol. 2014;15(1):203.
130. Re A, Joshi T, Kulberkyte E, Morris Q, Workman CT. RNA-protein interactions: an overview. Methods Mol Biol (Clifton, NJ) 2014;1097:491–521.
131. Zhang C, Darnell RB. Mapping in vivo protein-RNA interactions at single-nucleotide resolution from HITS-CLIP data. Nat Biotechnol. 2011;29(7):607–14.
132. Xiong HY, Barash Y, Frey BJ. Bayesian prediction of tissue-regulated splicing using RNA sequence and cellular context. Bioinformatics. 2011;27(18):2554–62.
133. Barbosa-Morais NL, Irimia M, Pan Q, Xiong HY, Gueroussov S, Lee LJ, Slobodeniuc V, Kutter C, Watt S, Colak R, et al. The evolutionary landscape of alternative splicing in vertebrate species. Science. 2012;338(6114):1587–93.
134. Busch A, Hertel KJ. Splicing predictions reliably classify different types of alternative splicing. RNA (New York, NY). 2015;21(5):813–23.
135. de Klerk E, t Hoen PA. Alternative mRNA transcription, processing, and translation: insights from RNA sequencing. Trends Genet: TIG. 2015;31(3):128–39.
136. Bentley DL. Coupling mRNA processing with transcription in time and space. Nat Rev Genet. 2014;15(3):163–75.
137. de Almeida SF, Carmo-Fonseca M. Reciprocal regulatory links between cotranscriptional splicing and chromatin. Semin Cell Dev Biol. 2014;32:2–10.

138. Zhou HL, Luo G, Wise JA, Lou H. Regulation of alternative splicing by local histone modifications: potential roles for RNA-guided mechanisms. Nucleic Acids Res. 2014; 42(2):701–13.
139. Rabani M, Raychowdhury R, Jovanovic M, Rooney M, Stumpo DJ, Pauli A, Hacohen N, Schier AF, Blackshear PJ, Friedman N, et al. High-resolution sequencing and modeling identifies distinct dynamic RNA regulatory strategies. Cell. 2014;159(7):1698–710.
140. Davis-Turak JC, Allison K, Shokhirev MN, Ponomarenko P, Tsimring LS, Glass CK, Johnson TL, Hoffmann A. Considering the kinetics of mRNA synthesis in the analysis of the genome and epigenome reveals determinants of co-transcriptional splicing. Nucleic Acids Res. 2015;43(2):699–707.
141. de Pretis S, Kress T, Morelli MJ, Melloni GE, Riva L, Amati B, Pelizzola M. INSPEcT: a computational tool to infer mRNA synthesis, processing and degradation dynamics from RNA-and 4sU-seq time course experiments. Bioinformatics 2015.
142. Nojima T, Gomes T, Grosso AR, Kimura H, Dye MJ, Dhir S, Carmo-Fonseca M, Proudfoot NJ. Mammalian NET-seq reveals genome-wide nascent transcription coupled to RNA processing. Cell. 2015;161(3):526–40.
143. Mayer A, di Iulio J, Maleri S, Eser U, Vierstra J, Reynolds A, Sandstrom R, Stamatoyannopoulos JA, Churchman LS. Native elongating transcript sequencing reveals human transcriptional activity at nucleotide resolution. Cell. 2015;161(3):541–54.
144. Chandola U, Das R, Panda B. Role of the N6-methyladenosine RNA mark in gene regulation and its implications on development and disease. Briefings Funct Genomics. 2015; 14(3):169–79.
145. Liu N, Dai Q, Zheng G, He C, Parisien M, Pan T. N(6)-methyladenosine-dependent RNA structural switches regulate RNA-protein interactions. Nature. 2015;518(7540):560–4.
146. Rieder LE, Reenan RA. The intricate relationship between RNA structure, editing, and splicing. Semin Cell Dev Biol. 2012;23(3):281–8.
147. Hube F, Francastel C. Mammalian introns: when the junk generates molecular diversity. Int J Mol Sci. 2015;16(3):4429–52.
148. Jeck WR, Sharpless NE. Detecting and characterizing circular RNAs. Nat Biotechnol. 2014;32(5):453–61.
149. Lasda E, Parker R. Circular RNAs: diversity of form and function. RNA. 2014;20(12): 1829–42.
150. Liang D, Wilusz JE. Short intronic repeat sequences facilitate circular RNA production. Genes Dev. 2014;28(20):2233–47.
151. Ivanov A, Memczak S, Wyler E, Torti F, Porath HT, Orejuela MR, Piechotta M, Levanon EY, Landthaler M, Dieterich C, et al. Analysis of intron sequences reveals hallmarks of circular RNA biogenesis in animals. Cell Rep. 2015;10(2):170–7.
152. Wang Y, Wang Z. Efficient backsplicing produces translatable circular mRNAs. RNA. 2015;21(2):172–9.
153. Ashwal-Fluss R, Meyer M, Pamudurti NR, Ivanov A, Bartok O, Hanan M, Evantal N, Memczak S, Rajewsky N, Kadener S. circRNA biogenesis competes with pre-mRNA splicing. Mol Cell. 2014;56(1):55–66.
154. Li Z, Huang C, Bao C, Chen L, Lin M, Wang X, Zhong G, Yu B, Hu W, Dai L, et al. Exon-intron circular RNAs regulate transcription in the nucleus. Nat Struct Mol Biol. 2015;22(3):256–64.
155. Conn SJ, Pillman KA, Toubia J, Conn VM, Salmanidis M, Phillips CA, Roslan S, Schreiber AW, Gregory PA, Goodall GJ. The RNA binding protein quaking regulates formation of circRNAs. Cell. 2015;160(6):1125–34.

156. Philips AV, Timchenko LT, Cooper TA. Disruption of splicing regulated by a CUG-binding protein in myotonic dystrophy. Science. 1998;280(5364):737–41.
157. Zhang J, Manley JL. Misregulation of pre-mRNA alternative splicing in cancer. Cancer Discovery. 2013;3(11):1228–37.
158. Brooks AN, Choi PS, de Waal L, Sharifnia T, Imielinski M, Saksena G, Pedamallu CS, Sivachenko A, Rosenberg M, Chmielecki J, et al. A pan-cancer analysis of transcriptome changes associated with somatic mutations in U2AF1 reveals commonly altered splicing events. PLoS ONE. 2014;9(1):e87361.
159. Dorman SN, Viner C, Rogan PK. Splicing mutation analysis reveals previously unrecognized pathways in lymph node-invasive breast cancer. Sci Rep. 2014;4:7063.
160. Irimia M, Weatheritt RJ, Ellis JD, Parikshak NN, Gonatopoulos-Pournatzis T, Babor M, Quesnel-Vallieres M, Tapial J, Raj B, O'Hanlon D, et al. A highly conserved program of neuronal microexons is misregulated in autistic brains. Cell. 2014;159(7):1511–23.
161. Ohe K, Hagiwara M. Modulation of alternative splicing with chemical compounds in new therapeutics for human diseases. ACS Chem Biol. 2015;10(4):914–24.
162. Svasti S, Suwanmanee T, Fucharoen S, Moulton HM, Nelson MH, Maeda N, Smithies O, Kole R. RNA repair restores hemoglobin expression in IVS2-654 thalassemic mice. Proc Natl Acad Sci USA. 2009;106(4):1205–10.
163. Arechavala-Gomeza V, Khoo B, Aartsma-Rus A. Splicing modulation therapy in the treatment of genetic diseases. Appl Clini Genet. 2014;7:245–52.
164. Wally V, Murauer EM, Bauer JW. Spliceosome-mediated trans-splicing: the therapeutic cut and paste. J Invest Dermatol. 2012;132(8):1959–66.
165. Havens MA, Duelli DM, Hastings ML. Targeting RNA splicing for disease therapy. Wiley Interdisc Rev RNA. 2013;4(3):247–66.
166. Kierlin-Duncan MN, Sullenger BA. Using 5′-PTMs to repair mutant beta-globin transcripts. RNA. 2007;13(8):1317–27.
167. Jiang H, Wong WH. Statistical inferences for isoform expression in RNA-Seq. Bioinformatics. 2009;25(8):1026–32.
168. Reyes-Herrera PH, Ficarra E. Computational Methods for CLIP-seq Data Processing. Bioinform Biol Insights. 2014;8:199–207.

Author Biographies

Yun-Hua Esther Hsiao graduated from the University of California, San Diego, with a B.S. degree in Bioengineering: Bioinformatics in 2012. She is currently a PhD student in Dr. Xinshu (Grace) Xiao's laboratory at UCLA. Her research focuses on the development and application of bioinformatic methods for high-throughput sequencing data analysis. Her main projects aim to better understand the regulation of alternative splicing and other aspects of post-transcriptional regulation.

Ashley Cass received her B.S. in Computational and Systems Biology at UCLA in 2011. She is currently pursuing a PhD in Bioinformatics at UCLA and is a member of Dr. Xinshu (Grace) Xiao's laboratory. Her research focus is in using and developing bioinformatic methods to analyze RNA regulation and degradation, particularly mechanisms and consequences of small RNA-mediated regulation.

Dr. Jae Hoon Bahn received his PhD degree in Comparative and Experimental Medicine at University of Tennessee, Knoxville, in 2010. He is currently a postdoctoral researcher working in Dr. Xinshu (Grace) Xiao's laboratory at UCLA. As a bench scientist, he is working on the molecular mechanisms of post-transcriptional gene regulation, encompassing a number of topics in splicing regulation and regulatory mechanisms of RNA editing. Dr. Bahn has published 43 scientific papers with an h-index of 18 and more than 1000 citations.

Dr. Xianzhi Lin received his diploma in Bioengineering from the Kunming University of Science and Technology in 2004 and a PhD in Microbiology from the Institute Pasteur of Shanghai, Chinese Academy of Sciences in 2012. He is currently a postdoctoral researcher working in Dr. Xinshu (Grace) Xiao's laboratory at UCLA. His research focuses on the molecular mechanisms of RNA regulation, including alternative splicing, RNA editing, and RNA degradation.

Dr. Xinshu (Grace) Xiao is an associate professor in Integrative Biology and Physiology and is a member of the Bioinformatics Inter-Departmental Program (IDP), the Jonsson Comprehensive Cancer Center, and the Molecular Biology Institute of UCLA. Dr. Xiao's research focuses on the bioinformatics and genomics of RNA biology. Work in the Xiao laboratory is highly interdisciplinary, bridging bioinformatics, genomics, systems biology, and basic molecular biology. Her laboratory has focused on the computational and experimental studies of alternative splicing and its regulation, RNA editing, and small RNA regulation of gene expression. Dr. Xiao's group developed a number of new bioinformatic methods for studies of RNA regulation, including multiple methods to accurately identify A-to-I RNA editing sites using RNA-Seq data with or without genome data, new methods to predict genetically regulated alternative splicing and polyadenylation events, and new short read aligners. The Xiao laboratory also applies existing and new methodologies to large-scale data analysis related to various diseases and biological processes, with the ultimate goal being an integrative and systematic understanding of RNA biology.

Chapter 3
Long Noncoding RNAs: Critical Regulators for Cell Lineage Commitment in the Central Nervous System

Xiaomin Dong, Naveen Reddy Muppani and Jiaqian Wu

Abstract Less than 3 % of the human genome encodes protein sequences and the majority of transcribed sequences are noncoding. Long noncoding RNAs (lncRNAs) refer to transcripts lacking in protein-coding potential and longer than 200 nt. lncRNAs are less conserved across species and expressed at a relatively lower level. The expression patterns of lncRNAs are more cell-type-specific than protein-coding genes. Currently, there are 8359 lncRNA genes annotated in Mouse GENCODE version M6 and 15,931 in Human GENCODE version 23. The number of lncRNA genes is still steadily increasing. Many lncRNAs have been shown to play crucial roles in regulating the expression of protein-coding genes during various biological processes. Particularly, lnRNAs can function as regulators during development and cell differentiation. Herein, we discussed the regulated expression and the functions of lncRNAs, as well as the underlying molecular mechanisms. Specifically, we highlighted the importance of lnRNAs in the central nervous system, and their regulatory roles during neural cell-fate determination.

Keywords Long noncoding RNAs · The central nervous system · Cell-fate determination · Transcriptional regulation

X. Dong · N.R. Muppani · J. Wu (✉)
The Vivian L. Smith Department of Neurosurgery, University of Texas Medical School at Houston, Houston, TX 77030, USA
e-mail: jiaqian.wu@uth.tmc.edu

X. Dong · N.R. Muppani · J. Wu
Center for Stem Cell and Regenerative Medicine, UT Brown Institution of Molecular Medicine, Houston, TX, USA

J. Wu, *Transcriptomics and Gene Regulation*,
Translational Bioinformatics 9, DOI 10.1007/978-94-017-7450-5_3

3.1 Introduction

Only a small portion of the mammalian genome encodes proteins [1–3]. Transcription of the genome is developmentally regulated and it generates a large number of noncoding RNAs, such as transfer RNAs (tRNAs), ribosomal RNAs (rRNAs), microRNAs (miRNA), and long noncoding RNAs (lncRNAs). tRNAs and rRNAs are important components of the protein synthesis machinery and play key roles in cellular protein synthesis. miRNAs and lncRNAs are considered to play crucial roles in regulation of gene expression [4, 5]. lncRNAs are nonprotein-coding transcripts and their size is longer than 200 bp. Thousands of lncRNAs were identified in mice and humans through RNA-sequencing studies [3, 6, 7]. lncRNAs were found to be controlled by various regulatory factors and displayed spatiotemporal expression profiles [8]. lncRNAs appeared to have functions in a variety of physiological processes such as development, homeostasis, stress response, and differentiation [3, 9, 10]. lncRNAs were shown to regulate various biological processes by influencing the expression of key molecules involved in these processes at transcriptional and posttranscriptional levels. Thus, lncRNAs and their functional diversity have received a great deal of interest in recent years.

Recent advances in genomic and molecular approaches provided novel insights and tools toward understanding the mechanisms underlying the function of lncRNAs [11, 12]. However, it is difficult to predict functions of lncRNAs across species due to poor sequence conservation. Many lncRNAs were found to modulate both individual gene expression and global gene networks in response to complex developmental and environmental signals. Recent studies demonstrated lncRNAs could act in *cis* to activate or silence neighboring protein-coding gene expression. Besides, lncRNAs could also exert *trans-effect* to regulate gene expression by interacting with various cellular factors. Although lncRNAs may act in *cis* or in *trans* to modulate single gene expression, the global changes in gene networks were mainly induced by *trans-effect* of lncRNAs. For the *trans-effect* of lncRNAs, emerging evidence indicates that lncRNAs could act as scaffolds to form functional complexes with epigenetic regulators and recruit these complexes at their action sites to affect the expression of individual gene or gene clusters by modulating chromatin epigenetic states [13]. In addition, lncRNAs physically interact with transcription factors and modulate their transcriptional activities on multiple target genes. Furthermore, interaction of lncRNAs with splicing factors was found to cause alternative splicing and affect global splicing patterns [14].

lncRNAs have emerged as important regulators in the central nervous system (CNS). The findings from the ABA (Allen Brain Atlas) data revealed that 849 out of 1328 lncRNAs examined were expressed within the adult mouse brain [8, 15]. Most of them were transcribed in a developmentally regulated and cell-type-specific manner. This suggested that lncRNAs were highly integrated into regulatory networks and played critical roles in differentiation of neural cells in the CNS. Additionally, perturbations of lncRNAs' expression were detected in neurological disorders, further supporting the possibility that lncRNAs are important regulatory factors in CNS [16–19]. In this chapter, we review recent emergent studies and highlight lncRNAs as important regulators for cell lineage commitment in CNS.

3.2 Identification and Classification of Long Noncoding RNA

Large-scale RNA-sequencing projects revealed that a majority of the mammalian genome is extensively transcribed and produces a large number of nonprotein-coding transcripts [1–3, 20–27]. Application of genome tiling array technology and deep sequencing to transcriptome profiling revealed that thousands of loci in mammals were transcribed to generate long nonprotein-coding transcripts [22, 28]. lncRNAs are one kind of these nonprotein-coding transcripts with a length ranging from 200 bp to several kilobases. lncRNAs were classified into various categories, such as intronic and exonic, intergenic, overlapping, bidirectional, and sense or antisense, relative to adjacent protein-coding genes (as shown in Fig. 3.1) [29–31]. Intergenic lncRNAs are transcribed from the regions between two protein-coding genes. Transcription of lncRNAs can also be initiated from an intron or exon of a gene. In some cases, a part of lncRNA sequences could be overlapping with the protein-coding sequences of a gene, although these lncRNAs do not have the capability of encoding any protein. Bidirectional lncRNAs are oriented head to head with an adjacent protein-coding gene, and the distance between their TSS (transcription start site) is less than 1 kb. lncRNAs are transcribed either in sense or antisense orientation compared to the direction of transcription of adjacent protein-coding genes.

lncRNAs may undergo alternative splicing and contain typical characteristics of mRNAs, such as 5′ 7-methylguanosine capping and 3′ polyadenylation [32–34]. A number of them exhibit specific temporal and spatial expression patterns in various tissues or cell types [3, 6, 8]. In mammalian genome, chromatin regions actively transcribed by RNA polymerase II to produce multi-exonic transcripts are marked by H3K4me3 and H3K36me3 chromatin modification-signatures (K4–K36 signature) [35]. Transcribing lncRNAs can also share these chromatin signatures indicating they may be produced through a similar biogenic process as mRNA [36]. Together, the nonprotein-coding transcripts transcribed from chromosome regions harboring the K4–K36 signature, containing single or multiple exons, and not completely overlapping with known protein-coding genes, could be considered as lncRNAs. These criteria have been used recently in several lncRNA discovery studies [37].

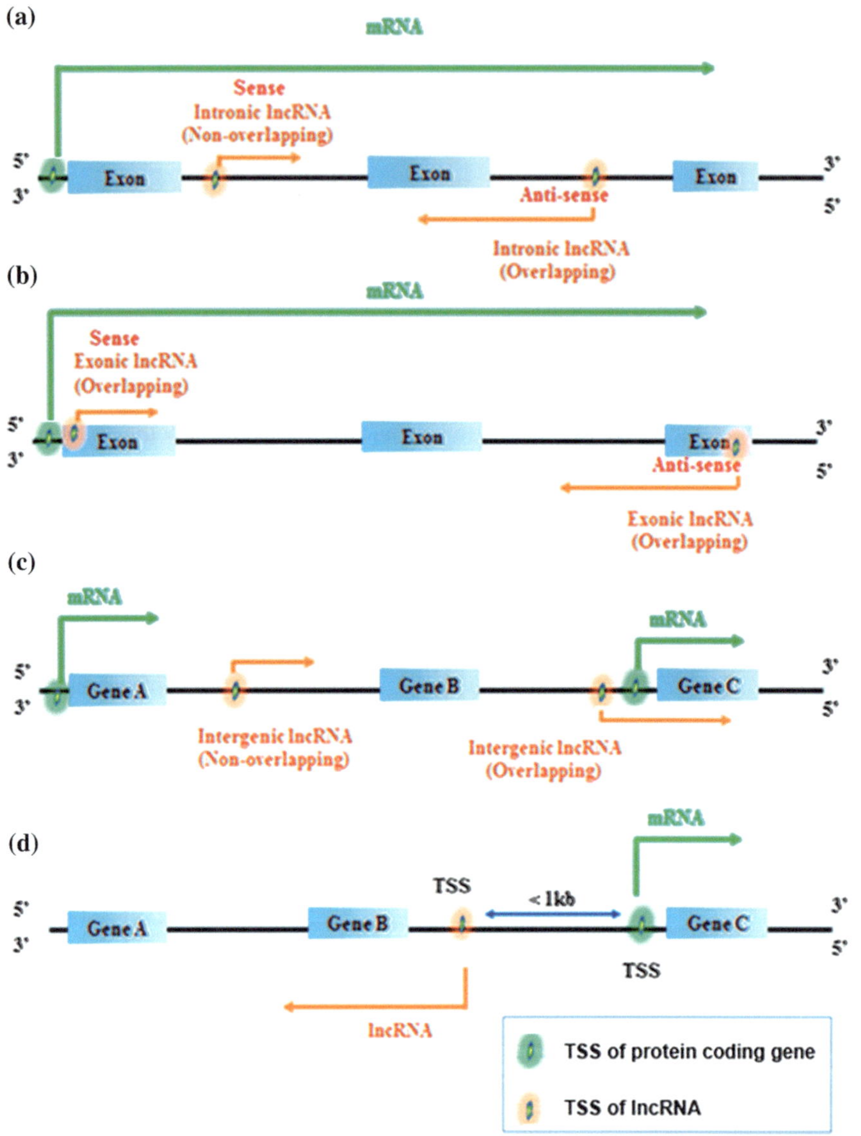

Fig. 3.1 Transcription of lncRNAs. lncRNAs transcription initiation site can be located within an intron or exon region of a protein-coding gene or regions between two different protein-coding genes on chromosomes. lncRNAs can be transcribed in either sense or antisense orientation (**a**, **b**). lncRNAs may locate between two different protein-coding genes. Some lncRNAs may overlap with a part of an exon of the adjacent protein-coding gene (**c**). Although they contain a part of an exon sequence of a known protein-coding gene, they fail to encode any protein. Transcription of lncRNA can also be initiated on the same point of DNA (less than 1 kb distance from adjacent protein-coding gene), but on the opposite strand of adjacent protein-coding gene (**d**)

3.3 Regulation of lncRNAs Expression

Developmental and tissue-specific temporal expression of lncRNAs suggested that their expression is tightly regulated via various regulatory mechanisms, such as transcription factors, epigenetic modifiers, and miRNA. lncRNAs were also shown to modulate the cellular levels of lncRNAs as shown in Fig. 3.2.

3.3.1 *Transcriptional Regulation of lncRNA Expression*

Transcription factors were found to regulate the expression of many lncRNAs [38]. Several lncRNAs contain putative binding sites for transcription factors, such as NF-κB, NANOG, OCT4, and SOX2 [39–41]. lncRNA *JADE* contains five putative NF-κB-binding sites. Upon DNA damage, NF-κB induces the *JADE* promoter activation [41]. p53 transactivates several lncRNAs, such as *lincRNA-p21* and *lincRNA-Mkln1* [42]. Apart from transcription factors regulation of lncRNAs, diverse epigenetic modifications might influence their transcription as well [43, 44].

Site-specific cytosine methylation was noticed in both *Xist* and *HOTAIR* within or near functionally important regions where chromatin-associated protein complexes can be recruited [45]. Up-regulation of various lncRNAs was observed as a result of decreased methylation at CpG islands of their promoter regions [46]. This

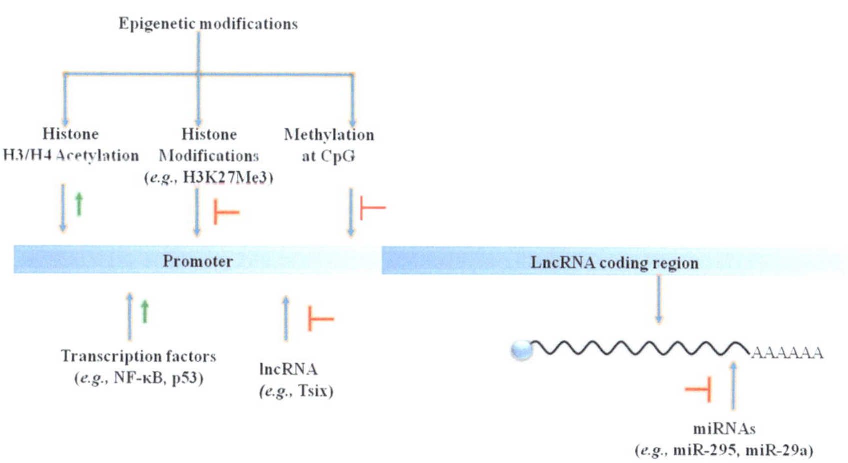

Fig. 3.2 Regulation of lncRNAs expression. lncRNAs expression could be regulated at transcriptional and posttranscriptional levels. Transcription factors, lncRNAs, methylation of CpG islands, and histone epigenetic modifications at lncRNA promoter region are the main mechanisms that regulate expression of lncRNA transcripts at the transcriptional level. Posttranscriptional regulation includes miRNA-mediated degradation of lncRNAs

suggested that methylation on cytosine may serve as a general mechanism in the regulation of expression of long noncoding RNAs.

In addition to DNA methylation, several studies reported that epigenetic modifications on histones can influence lncRNAs expression. For example, hypoxia-induced histone deacetylase 3 caused histone deacetylation at the promoter region of *lncRNA-LET*. This led to a down-regulation in *lncRNA-LET* expression [47]. When hepatocellular carcinoma (HCC) cells were treated with trichostatin A (TSA), a histone deacetylase inhibitor, lncRNA *uc002mbe.2*, was up-regulated [48]. EZH2, a histone-lysine N-methyltransferase, is the catalytic subunit of the Polycomb Repressive Complex 2 (PRC2). In embryonic stem cells, EZH2 mediates H3K27 methylation and thereby repressing the expression of several lncRNAs (e.g., *FT31040* and *FT38422*) [49]. In addition, some lncRNAs can be regulated by other lncRNAs. For example, *Xist* expression is blocked by *Tsix*, a lncRNA transcribed in the antisense orientation from a promoter downstream of *Xist* [50].

3.3.2 Posttranscriptional Regulation of lncRNA Expression

miRNA-mediated degradation is a common posttranscriptional mechanism that influences cellular mRNA levels. Similar to mRNAs, lncRNAs expression was found to be regulated by miRNAs. A direct regulation of lncRNA *MEG3* (maternally expressed gene 3) expression by the micro-RNA, miR-29a, was reported [43]. The miR-295 as a part of the Dicer-miRNA-Myc circuit promoted transcription of various lncRNAs [51]. Furthermore, the involvement of miRNA-671, miRNA-let-7b, miRNA-141, and miRNA-9 in degradation of different lncRNAs was also reported [52–54].

Thus, the transcriptional and posttranscriptional regulatory mechanisms modulating the expression of lncRNAs seem to be similar to most protein-coding genes. Developmental-specific expression of transcription factors, histone modifications at lncRNAs promoter regions, and miRNAs seem to play roles in the regulation of lncRNAs expression in a developmental stage specific manner.

3.4 Functions of lncRNAs

To date, the potential functions of most lncRNAs have not been well characterized. Initially, lncRNAs were thought to mainly serve as precursors for small RNAs [21]. Compared to protein-coding RNAs, lncRNAs are usually expressed at a lower level and lack apparent sequence conservation across species. These observations fueled the debate whether lncRNAs are generated due to transcriptional noise resulting from low RNA polymerase fidelity, or if they are artifacts of high-throughput sequencing [3, 55, 56]. However, studies in mice showed that the expression of many lncRNAs was developmental context restricted [29, 57]. Many of them were specifically expressed during embryonic stem cell differentiation and in the brain. These studies also revealed specific subcellular localization of lncRNAs [8, 58–60]. A recent study showed that lncRNAs were expressed in a more tissue-specific manner than protein-coding genes across 24 human tissues and cell types observed [3]. This evidence suggests that many, if not all, lncRNAs are under explicit control and may have critical functions during development [61].

Compared to miRNAs or protein-coding mRNAs, it is more difficult to infer lncRNAs function directly from their sequence or structure, because of the large diversity and poor sequence conservation across species. Identifying individual lncRNA function usually requires direct functional tests, e.g., loss-of-function or gain-of-function experiments [37]. An evolutionary analysis on lncRNAs revealed their conserved function despite limited sequence conservation [36, 62–64]. Several lines of research evidence indicate their involvement in a broad range of biological activities, such as chromosomal dynamics, telomere biology, and subcellular structural organization [61].

3.5 Molecular Mechanisms of lncRNAs

The precise mechanisms by which lncRNAs may function are not fully understood yet. lncRNAs may elicit their biological functions by interacting with protein complexes that transactivate genes or they may modify chromatin epigenetic states, and RNA editing, etc.

3.6 Involvement of lncRNAs in Transcriptional Regulation of Gene Expression

Enhancer and promoter regions of protein-coding genes are bound by transcription factors and cofactors. Interactions of lncRNA with these factors may participate in the process of regulation of transcription of various genes in *cis* or in *trans* [65, 66]. lncRNAs can be transcribed from enhancer or promoter regions of the protein-coding genes and recruit chromatin-binding proteins and modulate nearby protein-coding gene expression on the same chromosome in *cis*. For example, a lncRNA transcribed from the upstream of cyclin D1 binds to RNA-binding protein TLS (also known as FUS). This interaction is required for TLS to inhibit the histone acetyltransferase activities of CREB-binding protein and p300 to repress cyclin D1 expression [67]. In mouse, lncRNA *Evf-2* is transcribed from *Dlx5/6* locus and can recruit a transcription factor DLX2 at this locus. The *Evf-2* acts as a cofactor and promotes transcriptional activity of DLX2 to transactivate the adjacent genes at the *Dlx5/6* locus [68].

Additionally, lncRNAs can also regulate the transcription of target genes on other chromosomes in *trans*, without impacting their neighboring genes. Generally, these lncRNAs alter global transcription by influencing the activity of transcription factors in *trans*. One study demonstrated that upon depletion of nutrients or growth factors, a noncoding RNA named growth arrest-specific 5 (*Gas5*) was highly expressed in growth-arrested cells. *Gas5* was found to act as a glucocorticoid response element (GRE) and competed with other GRE to bind to the glucocorticoid receptor (GR). By preventing the binding of GR to correct regulatory GRE and repressing the transcriptional activity of GR, *Gas5* inhibited the expression of several responsive genes associated with cell survival and metabolism [69]. lncRNA *B2* was shown to mediate regulation of global gene transcription *in trans*. This lncRNA bound and suppressed RNA polymerase II activity. *B2* was induced upon a stress condition, and its induction caused a global transcription repression [70, 71].

lncRNAs can act as either activators or repressors by influencing epigenetic processes in *cis* or in *trans*. Functional involvements of lncRNAs in chromatin modifications were also reported [72]. Various studies revealed that lncRNAs could assist epigenetic changes by recruiting chromatin remodeling complexes to specific genomic loci [61, 73]. For example, *Xist* (inactive X-specific transcript, causes X chromosome inactivation in *cis*) and *HOTAIR* (HOX transcript antisense RNA, silences HOXD locus in *trans*) mediated recruitment of PRC2 on their respective targets [74, 75]. A study showed that different regions of *HOTAIR* bind to different subunits of the histone modification complexes, functioning as a modular scaffold for chromatin-modifying complex formation to regulate expression of various genes [76]. The lncRNA *Air* is antisense to *Igf2r* (insulin-like growth-factor type-2 receptor) gene [77]. It promotes the silencing of *Igf2r, Slc22a2 and Slc22a3* by triggering G9a methyltransferase recruitment at the promoter region of these genes [78]. The lncRNA *Kcnq1ot1* (*Kcnq1* opposite transcript1) was transcribed from intron 11 of the *Kcnq1* gene in the antisense direction [79]. It regulates methylation of the *Kcnq1* gene promoter by interacting with DNMT1 (DNA methyltransferase 1) and recruiting EZH2, as well as G9a at the *Kcnq1* gene promoter [80–83]. In addition to these early examples, several other lncRNAs were also found to be associated with chromatin-modifying complexes and affect gene expression [73]. In humans, ~20 % lncRNAs bind to the PRC complex, and 52 % of all known lncRNAs can interact with other chromatin-modifying complexes such as CoREST and SMCX [73]. These findings evidenced that lncRNAs can bind to chromatin-modifying complexes to guide them to specific locations on the chromosome. Differentially expressed lncRNAs can bind to several ubiquitously expressed chromatin-modifying complexes and thereby helping to establish cell type and condition-specific epigenetic states [61]. Such a mechanism resolves the paradox that a small repertoire of chromatin-modifying complexes, which are often comprised of RNA-binding domains with little DNA sequence specificity, were able to establish complex epigenetic states across different cell types arising throughout development [61]. The ability of lncRNAs to recruit chromatin-binding proteins (transcription factors or epigenetic modifiers) to gene promoters can largely

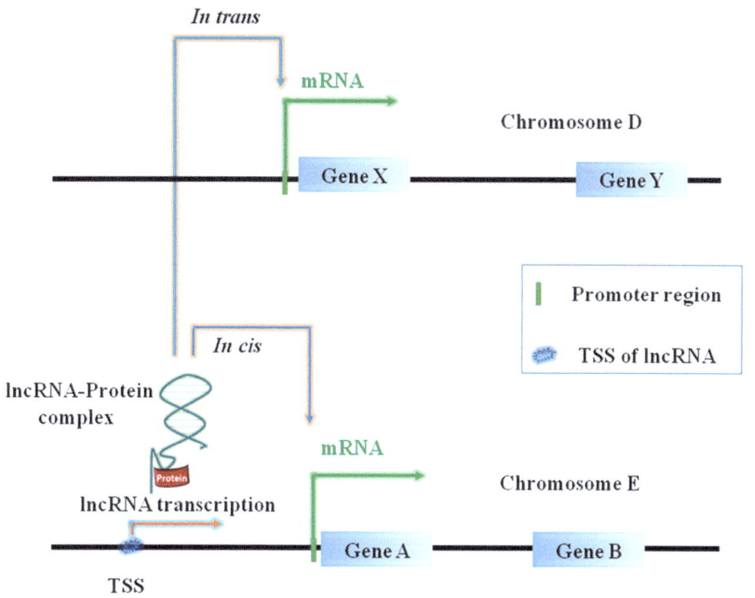

Fig. 3.3 Involvement of lncRNAs in transcriptional regulation of gene expression. lncRNAs could regulate gene expression in *cis* or in *trans*. They form complexes with transcription factors or epigenetic modifiers and thereby influencing the transcription of genomic adjacent protein-coding genes on the same chromosome in *cis* or genes on the other chromosome in *trans*

expand their transcriptional regulatory repertoire, as different protein factors can be combined together with various lncRNAs to regulate specific target gene expression as shown in Fig. 3.3.

3.6.1 Involvement of lncRNAs in Posttranscriptional Regulation of Gene Expression

miRNAs can influence mRNA decay by binding to 3′ UTR (untranslated region) of their target genes, thus affecting protein translation. Similar to miRNAs, some lncRNAs also have the ability to recognize complementary sequences which allow highly specific interactions, and bind to their target mRNAs. These interactions regulate posttranscriptional modifications of mRNAs, such as splicing, editing, transport, translation, and degradation as shown in Fig. 3.4 [61]. Many lncRNAs were transcribed as antisense transcripts. An example is the *zeb2nat* (Zeb2 antisense RNA), which complements the 5′ splice site of an intron in *Zeb2* 5′ UTR. The expression of *zeb2nat* prevented the splicing of this intron required for proper translation of the ZEB2 protein. Similar to *zeb2nat*, many other lncRNAs may also have regulatory functions in alternative splicing of mRNA. In addition, lncRNAs

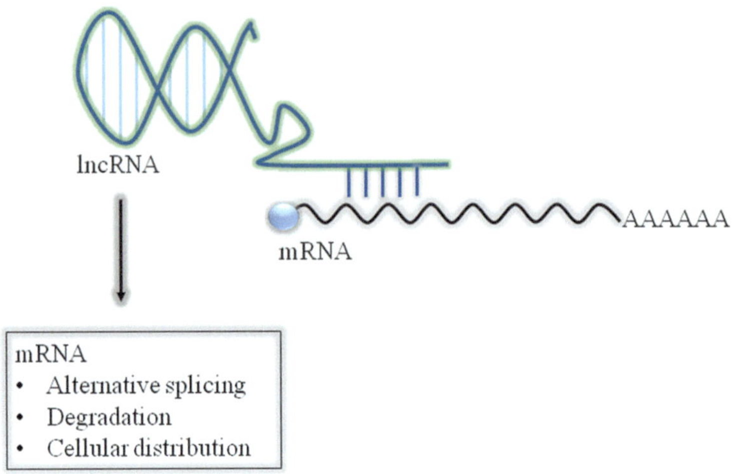

Fig. 3.4 Involvement of lncRNAs in posttranscriptional regulation of gene expression. lncRNAs were involved in posttranscriptional regulation of gene expression by modulating splicing, editing, transport, translation, and degradation of a mRNA

were also found to interact with splicing factors. For example, lncRNA *Malat1* (metastasis associated lung adenocarcinoma transcript 1) co-localized with many splicing factors in nuclear speckles [14]. Serine Arginine (SR)-splicing factors are capable of regulating alternative splicing and this capability depends on their concentration and phosphorylation. *Malat1* physically interacts with SR proteins. This interaction modulates the levels of phosphorylation on SR proteins, their cellular distribution and thereby influencing alternative splicing mediated by them.

Interestingly, lncRNAs could also compete with miRNAs for their binding site on target mRNAs. lncRNA *linc-RoR* was reported to play an important role in embryonic stem cell renewal. Since *linc-RoR* was highly expressed in embryonic stem cell, it titrated away microR-145 and protected OCT4, SOX2, and NANOG that are required for embryonic stem cell renewal [84]. lncRNAs were also found to influence protein subcellular localization. For example, the lncRNA *NRON* bound to the members of the nucleocytoplasmic trafficking machinery and specifically inhibited nuclear accumulation of the transcription factor NFAT and thereby its transcriptional activities [85]. The involvement lncRNAs in RNA editing, and their influence on localization of cellular factors explains their regulatory roles related to functions of proteins and several biological processes.

The above-mentioned evidence indicates that the lncRNA-mediated regulation of the expression of several genes is not only at the transcriptional level, but also at the posttranscriptional level.

3.7 Expression and Function of lncRNAs in CNS

The functions of lncRNAs in CNS were mainly investigated by analyzing their developmental, tissue- and cell-type-specific expression. Studies from the ABA (Allen Brain Atlas) data revealed that about two-thirds of lncRNAs were expressed within the adult mouse brain [8]. Expression of lncRNAs was found to be developmental stage and tissue-specific. In additional studies, gene expression analysis of both protein-coding and noncoding genes was performed in mouse brain and six layers of mouse neocortex by using in situ hybridization and RNA-Seq technology [15]. The results indicated that lncRNAs expression was significantly associated with genomic sequence nearby or overlapping protein-coding genes. Interestingly, the majority of these protein-coding genes were transcriptional regulators or factors involved in nervous system development.

In addition to studies in adult brain, the expression analyses of lncRNAs were also performed in embryonic brain [86, 87]. A recent study used next-generation sequencing to analyze lncRNAs expression in proliferating neural stem cells, differentiating progenitors and neurons [15]. Several lncRNAs were found to be involved in neurogenic commitment and neuronal survival [88]. Another study focused on the expression of lncRNAs during different stages of embryonic brain development [89]. This study revealed that thousands of embryonic, brain-specific lncRNAs are differentially expressed during brain development suggesting their potential involvement in brain development. Most of them are expressed at lower levels in adult brain compared to embryonic brain. Additionally, a large fraction of these lncRNAs located within imprinted gene (the expression of imprinted gene only occurs from one allele and is determined in a parent-specific manner) clusters was suggested to act in imprinting lncRNAs to control brain development, similar to other imprinted transcripts [88]. Taken together, this evidence implies that lncRNAs may play critical roles in controlling brain development.

Since dynamic changes of lncRNAs were observed during neural cell differentiation, it was suggested that lncRNAs expression is also regulated by cellular factors [90]. Both transcription factors and epigenetic modifiers were found to contribute to the modulation of lncRNAs expression in CNS.

The repressor element 1 silencing transcription factor (REST) functions as an important transcriptional repressor to control neuronal gene expression during nervous system development. Based on unbiased genome annotation in mouse and humans, about one-fourth of REST-binding sites were located within 10 kb of lncRNAs [91]. lncRNA *AK046052* and *AK090153* were identified to be REST target genes and their expression was inhibited by REST in neural stem cells.

The lncRNA rhabdomyosarcoma 2 associated transcript (*RMST*) was specifically expressed in the brain. Previous studies indicated *RMST* was regulated by transcription factor PAX2 at the mouse mid-hindbrain boundary [92]. More recently, binding of REST to *RMST* was revealed by ENCODE transcription factor binding analysis and ChIP experiments [93]. Moreover, REST knockdown resulted in increased expression of *RMST*, confirming the repressive function of REST on *RMST* transcription.

Another example is *utNgn1*-mediated epigenetic silencing of *Neurog1*. *utNgn1* is transcribed from one of the enhancer elements located 7 kb upstream of the *Neurog1* gene. The involvement of *Neurog1* in neuronal fate commitment has been described [94]. The expression of *utNgn1* was positively associated with *Neurog1* mRNA expression during neural development. Polycomb group (PcG) proteins are epigenetic chromatin modifiers. They can modify methylation states of histones at promoter regions and repress gene transcription. They were found to inhibit both the *utNgn1* and *Neurog1* expressions. On the other hand, Wnt3a was shown to up-regulate both *utNgn1* and *Neurog1* expressions [95].

3.8 Roles of lncRNAs in Cell Lineage Commitment in CNS

Neural stem cells (NSCs) are multipotent stem cells that have limited ability to differentiate compared to embryonic stem cells (ESCs) [96]. These cells exist in both the embryonic cortical ventricular zone (VZ) and postnatal ventricular–subventricular zone (V-SVZ) in the brain. They can both self-renew and differentiate to neurons, astrocytes, and oligodendrocytes. The proper generation of these cell types is of importance for the normal brain development. Disturbing this process may cause developmental defects or diseases in the brain [97]. In addition, NSCs can give rise to neurons as well as glial cells to facilitate the recovery process after brain injury, such as traumatic brain injury (TBI) and ischemic brain injury [98–100].

lncRNAs appear to be involved in the cell-type specification during NSCs differentiation. They were found to have cell-type and subcellular-specific expression patterns in brain [88]. Moreover, dramatic changes in lncRNAs expression were observed during neuronal–glial cell-fate switch, as well as neuronal and oligodendrocyte lineage differentiation in a previous study [90]. This study also noted that lncRNAs are located close to, and showed similar expression profiles, as protein-coding genes involved in the neural differentiation processes. Furthermore, another study demonstrated that lncRNAs *Dlx1as* and *Six3os* can regulate neurogenesis from NSCs. In this study, the expression of lncRNAs in the main cell types of the SVZ was analyzed and showed differential lnRNAs expression in neural cells types [101]. In addition, both up-regulated and down-regulated lncRNAs were identified during in vitro differentiation of neural stem cells. Knockdown studies to detect the roles of lncRNAs in neuronal differentiation were also performed and revealed that ablation of *Six3os* resulted in an increase in the number of astrocytes and a decrease in the number of neuron, as well as oligodendrocytes in neural stem cell differentiation. An increase of astrocytes and a decrease of neurons were also observed as a result of *Dlx1as* knockdown.

De novo identification of novel lncRNAs from RNA-Seq data of purified cell types can further expand the lncRNA repertoire and elucidate lncRNAs' function in cell lineage commitment in CNS. We reported 811 lncRNAs with the criterion of

FPKM > 1 from the eight brain cell types (neurons, astrocytes, oligodendrocyte precursor cells, newly formed oligodendrocytes, myelinating oligodendrocytes, microglia, endothelial cells, and pericytes from mouse cerebral cortex) based on the GENCODE annotation [27]. Some lncRNAs are highly expressed, with 12 having FPKM values >100 in the brain. We found that some lncRNAs are cell-type-specific or enriched. Astrocytes and neurons express larger numbers of lncRNAs (109 ± 2 and 92 ± 3, respectively) than myelinating oligodendrocytes and endothelial cells (44 ± 3 and 48 ± 3, respectively). We are in the process of broadening lncRNA catalog by ab initio transcriptome reconstruction and characterizing the regulation of unannotated lncRNAs in cell-fate determination by integrating differential gene expression and transcription factor occupancy information. lncRNAs involved in oligodendrocyte precursor differentiation from NSCs have been identified by loss-of-function experiments. All the above observations suggested a role of lineage-enriched lncRNAs in the cell-fate decision of NSCs. Some examples of lncRNAs involved in cell lineage commitment in CNS are shown in Table 3.1.

3.8.1 Cis-Acting *lncRNAs in Cell Lineage Commitment in CNS*

Attention had been focused on understanding the influence of lncRNAs in the transcription regulation of adjacent protein-coding genes on the same chromosome. Recently, a number of lncRNAs have been shown to function locally and control the expression of their neighboring genes to influence cell-fate specification and differentiation of CNS.

NKX2.2 is a known transcriptional factor regulating oligodendrocyte lineage differentiation [102]. *Nkx2.2 antisense* is transcribed in the antisense orientation to the *Nkx2.2* gene and localized in the cytoplasm. Overexpression of *Nkx2.2 antisense* in neural stem cells could act in *cis* to up-regulate *Nkx2.2* mRNA levels and promote oligodendrocyte lineage differentiation [103].

Evf-2 is another lncRNA acting as a transcriptional co-activator to regulate gene expression and cell-fate specification in CNS. It is expressed in immature neurons and its coding sequence is located in the intergenic region of the *Dlx5* and *Dlx6* genes. The DLX homeodomain transcription factors are known to play an important role in neural differentiation [115–118]. Ablation of *Dlx1*, *Dlx2*, *Dlx5*, and *Dlx6* led to a reduction in the number of interneurons in mice. Moreover, the DLX2 protein was demonstrated to bind to the *Dlx5/6* intergenic sequence region [119]. *Evf-2* was found to not only enhance and stabilize the binding of DLX2 protein to the *Dlx5/6* intergenic sequence [68], but also recruit DLX and MECP2 to the *Dlx5/6* intergenic region to inhibit the transcription of *Dlx-5* and *Dlx-6* [104]. In addition, loss of *Evf-2* function caused decreased GABAergic interneurons in postnatal day 2 hippocampus and dentate gyrus [104]. These multiple lines of evidence implicate a potential role for *Evf-2* in neural differentiation.

Table 3.1 Examples of lncRNA involved in cell lineage commitment in CNS

lncRNA	Classification	Function	Classification of effect	Ref.
Nkx2.2 antisense	Antisense	Up-regulates *Nkx2.2* mRNA levels and promote oligodendrocyte lineage differentiation	In *cis*	[103]
Evf-2	Intergenic	Recruits DLX and MECP2 to the *Dlx5/6* intergenic region to inhibit the transcription of *Dlx-5* and *Dlx-6*	In *cis*	[104]
Sox2ot	Overlapping	Up-regulates *Sox2* expression and may be functionally associated with Sox2 in neurogenesis	In *cis*	[105, 106]
utNgn1	Intergenic	Promotes the expression of *Neurog1* and *Tbr2*	In *cis*	[95]
RMST	Intergenic	Forms RNA-protein complex with SOX2 to regulate a set of downstream genes implicated in neurogenesis	In *trans*	[93]
RNCR2	Intergenic	Modulates retinal differentiation by interacting with SF1 splicing factor and affecting alternative splicing patterns of other genes	In *trans*	[107]
Tug1	Intergenic	Causes global changes in expression of photoreceptor genes during retinal development	In *trans*	[108]
Six3	Antisense	Affects the expression of Six3-associated genes through interacting with EZH2 during retinal development	In *trans*	[109, 110]
TUNA	Intergenic	Recruits PTBP1, hnRNP-K, and NCL to the Sox2 promoter region	In *trans*	[111]
Pnky	Intergenic	Interacts with polypyrimidine-tract-binding protein 1 (PTBP1) to regulate neurogenesis	In *trans*	[112]
Paupar	Intergenic	Regulates the transcription of *Pax6* locally and transcription of neuro-developmental genes distally	In *trans* and In *cis*	[113]
Dali	Intergenic	Regulates the transcription of *Pou3f3* locally and interact with the POU3F3 protein to regulate transcription of neural differentiation genes distally	In *trans* and In *cis*	[114]

Transcription factor SOX2 is required for NSCs pluripotency and neurogenesis. Its expression level is high in NSCs and down-regulated after NSCs differentiation [120]. Loss of SOX2 led to a reduced number of mature neurons after NSCs differentiation in vitro. Furthermore, it also caused a loss of GABAergic neurons and hippocampal neurogenesis in vivo [121]. *Sox2* gene is located in an intronic region of lncRNA *Sox2ot* (Sox2 overlapping transcript). *Sox2ot* is transcribed in the same direction as *Sox2* and has several isoforms, such as *Sox2ot-s1* and *Sox2ot-S2*

[105, 106]. With similar expression patterns to *Sox2*, *Sox2ot* was detected in cultures of mouse neurospheres in vitro and adult mouse tissues [8, 106]. More recently, a study showed that ectopic expression of *Sox2ot* up-regulated the expression of *Sox2* by 20-fold [122]. Thus, *Sox2ot* was suggested to regulate *Sox2* expression and may be functionally associated with *Sox2* in neurogenesis [105].

3.8.2 Trans-Acting *lncRNAs in Cell Lineage Commitment in CNS*

lncRNAs were also able to interact with epigenetic modifiers, transcription factors, and splicing factors and influence their molecular functions. Therefore, lncRNAs can distally target different loci in the genome to modulate global gene expression [59]. Recently, some *trans-acting* lncRNAs were shown to control lineage differentiation in the CNS.

The rhabdomyosarcoma 2 associated transcript (*RMST*) was found to be express mainly in the CNS, especially in the developing dopaminergic neurons of the ventral midbrain [123]. A previous study demonstrated that *RMST* expression was elevated during neuronal differentiation [93]. Depletion of *RMST* inhibited neurogenic program, while overexpression of *RMST* induced neuronal differentiation. This evidence implies a functional role of *RMST* in neural cell-fate decision [124]. Further investigations revealed that *RMST* was transcriptionally repressed by REST and interacted with the SOX2 protein. During neuronal differentiation, REST levels decrease [125, 126]. Meanwhile, up-regulated *RMST* may form RNA-protein complex with SOX2 to regulate a set of downstream genes implicated in neurogenesis [93].

Additionally, retinal noncoding RNA 2 (*RNCR2*) (also known as *Gomafu* or *Miat*) is located in the nucleus and is highly expressed in the developing retina. Furthermore, *RNCR2* was detected in differentiating oligodendrocytes, as well as neurons and exhibited a specific expression pattern in brain [8, 90]. In vivo electroporation was performed in mouse retina to knockdown the expression of *RNCR2*. The loss of *RNCR2* promoted retinal cell differentiation. *RNCR2* was exclusively retained in the nucleus as many other lncRNAs are and did not shuttle between the nucleus and cytoplasm [21, 127, 128]. When *RNCR2* was fused with IRES-GFP, the IRES-GFP induced translocation of *RNCR2* from the nucleus to the cytoplasm. This led to an enhancement in amacrine and Müller glial differentiation as caused by the loss of *RNCR2* in the nucleus [127]. These results suggest that nuclear-retained *RNCR2* regulated the retinal cell fate. Moreover, it was proposed that *RNCR2* may modulate retinal differentiation by interacting with SF1 splicing factor and affecting alternative splicing patterns of other genes [107].

Tug1 (taurine up-regulated gene 1) is another lncRNA involved in retinal development. *Tug1* was required for retinal differentiation and acted in *trans* to modulate photoreceptor genes through interactions with PRC2 [73]. Knockdown of

Tug1 caused global changes in expression of photoreceptor genes, improper photoreceptor development, and cell death [108].

Six3 (sine oculis homeobox homolog 3) was found to be expressed in the mouse retina and associated with retinal development [109]. *Six3OS* is the antisense-transcript of the *Six3* gene. To evaluate the effect of *Six3OS* on retinal development, both overexpression and knockdown experiments were performed in vivo [110]. The results indicated the involvement of *Six3OS* in retinal cell differentiation and that *Six3OS* may act in *trans* to affect the expression of *Six3*-associated genes through interactions with EZH2.

The highly conserved lncRNA, *TUNA* (Tcl1 upstream neuron-associated lincRNA) was found to be associated with neural development and function in zebrafish, mice, and humans [111]. It also plays a key role in pluripotency and neural differentiation of mouse ESCs. It is required for neural lineage commitment of ESCs and to maintain stemness of ESCs. According to RNA pulldown and RNA immunoprecipitation experiments, *TUNA* was found to interact with three multifunctional proteins, including polypyrimidine-tract-binding protein (PTBP1/PTB/hnRNP-I), heterogeneous nuclear ribonucleoprotein K (hnRNP-K), and nucleolin (NCL). Impact of NCL on chromatin remodeling and transcription was previously reported [129, 130]. Notably, hnRNP-K and PTBP-1 were implicated in neural differentiation through posttranscriptional mechanisms [131, 132]. *TUNA* recruited PTBP1, hnRNP-K, and NCL to the *Sox2* promoter region and promoted its transactivation. Moreover, *TUNA* knockdown inhibited neural lineage differentiation from ESC. Apart from their individual target genes, *TUNA* and *Sox2* may co-regulate a set of common genes to influence the ESC state and neurogenesis [111].

A neural-specific lncRNA Pinky (*Pnky*) is expressed in NSCs of the developing mouse and human brain [112]. It regulated neurogenesis from NSCs in the embryonic and postnatal brain. Knockdown of *Pnky* potentiated neuronal lineage commitment both in culture and in vivo. Its knockdown increased neuron production up to 4-fold in postnatal NSCs. Knockdown of *Pnky* in the embryonic mouse cortex led to an increase in neuronal differentiation and a decrease in the NSC population [112]. *Pnky* was shown to interact with Polypyrimidine-tract-binding protein 1 (PTBP1). PTBP1 is a RNA-splicing factor and can act as a repressor for neuronal differentiation. The expression of PTBP1 was decreased during neurogenesis and PTBP1 knockdown resulted in an increase in neurogenesis [133]. Furthermore, both *Pnky* and PTBP1 were found to regulate a common set of transcripts involved in neuronal differentiation [112].

3.8.3 lncRNAs with Both Cis *and* Trans-Effects *on Cell Lineage Commitment in CNS*

Some lncRNAs are able to function locally on nearby genes, as well as on distal genes located on different chromosomes. For example, lncRNA *CTBP1-AS* was shown to influence prostate cancer progression either by directly inhibiting the expression of C-terminal binding protein1 (CTBP1) in *cis* or by suppressing tumor-repressor genes globally in *trans* [134]. Recent studies demonstrated some lncRNAs with both *cis* and *trans-acting* function in lineage differentiation in the CNS.

An evolutionarily conserved lncRNA *Paupar* is transcribed from 8.5 kb upstream of *Pax6* gene [113]. Knockdown of *Paupar* was shown to control the growth and differentiation of neural cells by a loss-of-function assay. *Paupar* was able to locally regulate the transcription of *Pax6*. Transcription factor PAX6 was implicated in controlling progenitor cell potency and neuronal cell specification [135]. Additionally, *Paupar* was shown to bind to specific genomic regions distally and regulate the transcription of neuro-developmental genes *in trans* [113].

DNMT1-associated long intergenic noncoding RNA (*Dali*) is a conserved intergenic lncRNA expressed in the central nervous system [114]. It is transcribed from downstream of the *Pou3f3* transcription factor gene and co-expressed with *Pou3f3* in neural cell lineages. *Dali* regulates the transcription of the *Pou3f3* locus. Distally, it targets active promoters of neural differentiation genes and regulates their expression. This regulation was in part mediated through physically interacting with the POU3F3 protein. *Dali* epigenetically regulates neural differentiation. In mice and humans, *Dali* interacts with DNMT1 and regulates DNA methylation of CpG island-associated promoters in *trans*. Depletion of *Dali* disrupted differentiation of neuroblastoma cells [114].

3.9 Conclusions

In summary, we reviewed the significant roles of lncRNAs, especially their role as cell-fate determinants in CNS. However, the elucidation of the regulatory functions of lncRNAs in CNS cell-fate determination is limited so far. The cell lineage commitment could involve more complex regulatory mechanisms. Other functions of lncRNAs may exist and need to be discovered. Most of the previous studies employed lncRNAs overexpression or silencing methods to study functions of lncRNAs. As many lncRNAs are localized in the nucleus, the RNAi method might not efficiently knockdown their expression due to a lack of the RNAi machinery in the nucleus. Therefore, new technical advancements are highly desirable in order to gain further insights into the roles of lncRNAs in CNS development and other biological processes.

Acknowledgements We thank Dr. Eva Zsigmond for reading and editing our manuscript. JQW, XD, and NRM are supported by grants from the National Institutes of Health R01 NS088353 and R00 HL093213, the Staman Ogilvie Fund—Memorial Hermann Foundation, Mission Connect—a program of the TIRR Foundation, the Senator Lloyd & B. A. Bentsen Center for Stroke Research, UTHealth BRAIN Initiative and CTSA UL1 TR000371, and a grant from the University of Texas System Neuroscience and Neurotechnology Research Institute (Grant #362469).

References

1. Lee JT. Epigenetic regulation by long noncoding RNAs. Science. 2012;338(6113):1435–9.
2. Birney E, Stamatoyannopoulos JA, Dutta A, Guigo R, Gingeras TR, Margulies EH, et al. Identification and analysis of functional elements in 1 % of the human genome by the ENCODE pilot project. Nature. 2007;447(7146):799–816.
3. Cabili MN, Trapnell C, Goff L, Koziol M, Tazon-Vega B, Regev A, et al. Integrative annotation of human large intergenic noncoding RNAs reveals global properties and specific subclasses. Genes Dev. 2011;25(18):1915–27.
4. Ameres SL, Zamore PD. Diversifying microRNA sequence and function. Nat Rev Mol Cell Biol. 2013;14(8):475–88.
5. Geisler S, Coller J. RNA in unexpected places: long non-coding RNA functions in diverse cellular contexts. Nat Rev Mol Cell Biol. 2013;14(11):699–712.
6. Guttman M, Garber M, Levin JZ, Donaghey J, Robinson J, Adiconis X, et al. Ab initio reconstruction of cell type-specific transcriptomes in mouse reveals the conserved multi-exonic structure of lincRNAs. Nat Biotechnol. 2010;28(5):503–10.
7. Khorkova O, Myers AJ, Hsiao J, Wahlestedt C. Natural antisense transcripts. Hum Mol Genet. 2014;23(R1):R54–63.
8. Mercer TR, Dinger ME, Sunkin SM, Mehler MF, Mattick JS. Specific expression of long noncoding RNAs in the mouse brain. Proc Natl Acad Sci USA. 2008;105(2):716–21.
9. Tripathi V, Shen Z, Chakraborty A, Giri S, Freier SM, Wu X, et al. Long noncoding RNA MALAT1 controls cell cycle progression by regulating the expression of oncogenic transcription factor B-MYB. PLoS Genet. 2013;9(3):e1003368.
10. Kretz M, Siprashvili Z, Chu C, Webster DE, Zehnder A, Qu K, et al. Control of somatic tissue differentiation by the long non-coding RNA TINCR. Nature. 2013;493(7431):231–5.
11. Gong S, Zheng C, Doughty ML, Losos K, Didkovsky N, Schambra UB, et al. A gene expression atlas of the central nervous system based on bacterial artificial chromosomes. Nature. 2003;425(6961):917–25.
12. Morris JA, Royall JJ, Bertagnolli D, Boe AF, Burnell JJ, Byrnes EJ, et al. Divergent and nonuniform gene expression patterns in mouse brain. Proc Natl Acad Sci USA. 2010;107(44):19049–54.
13. Peschansky VJ, Wahlestedt C. Non-coding RNAs as direct and indirect modulators of epigenetic regulation. Epigenetics. 2014;9(1):3–12.
14. Tripathi V, Ellis JD, Shen Z, Song DY, Pan Q, Watt AT, et al. The nuclear-retained noncoding RNA MALAT1 regulates alternative splicing by modulating SR splicing factor phosphorylation. Mol Cell. 2010;39(6):925–38.
15. Lein ES, Hawrylycz MJ, Ao N, Ayres M, Bensinger A, Bernard A, et al. Genome-wide atlas of gene expression in the adult mouse brain. Nature. 2007;445(7124):168–76.
16. Massone S, Vassallo I, Fiorino G, Castelnuovo M, Barbieri F, Borghi R, et al. 17A, a novel non-coding RNA, regulates GABA B alternative splicing and signaling in response to inflammatory stimuli and in Alzheimer disease. Neurobiol Dis. 2011;41(2):308–17.
17. Soreq L, Guffanti A, Salomonis N, Simchovitz A, Israel Z, Bergman H, et al. Long non-coding RNA and alternative splicing modulations in Parkinson's leukocytes identified by RNA sequencing. PLoS Comput Biol. 2014;10(3):e1003517.
18. Johnson R. Long non-coding RNAs in Huntington's disease neurodegeneration. Neurobiol Dis. 2012;46(2):245–54.
19. Lin M, Pedrosa E, Shah A, Hrabovsky A, Maqbool S, Zheng D, et al. RNA-Seq of human neurons derived from iPS cells reveals candidate long non-coding RNAs involved in neurogenesis and neuropsychiatric disorders. PLoS ONE. 2011;6(9):e23356.
20. Carninci P, Kasukawa T, Katayama S, Gough J, Frith MC, Maeda N, et al. The transcriptional landscape of the mammalian genome. Science. 2005;309(5740):1559–63.

21. Kapranov P, Cheng J, Dike S, Nix DA, Duttagupta R, Willingham AT, et al. RNA maps reveal new RNA classes and a possible function for pervasive transcription. Science. 2007;316(5830):1484–8.
22. Okazaki Y, Furuno M, Kasukawa T, Adachi J, Bono H, Kondo S, et al. Analysis of the mouse transcriptome based on functional annotation of 60,770 full-length cDNAs. Nature. 2002;420(6915):563–73.
23. Wu JQ, Du J, Rozowsky J, Zhang Z, Urban AE, Euskirchen G, et al. Systematic analysis of transcribed loci in ENCODE regions using RACE sequencing reveals extensive transcription in the human genome. Genome Biol. 2008;9(1):R3.
24. Wu JQ, Habegger L, Noisa P, Szekely A, Qiu C, Hutchison S, et al. Dynamic transcriptomes during neural differentiation of human embryonic stem cells revealed by short, long, and paired-end sequencing. Proc Natl Acad Sci USA. 2010;107(11):5254–9.
25. Rozowsky J, Wu J, Lian Z, Nagalakshmi U, Korbel JO, Kapranov P, et al. Novel transcribed regions in the human genome. Cold Spring Harb Symp Quant Biol. 2006;71:111–6.
26. Bertone P, Stolc V, Royce TE, Rozowsky JS, Urban AE, Zhu X, et al. Global identification of human transcribed sequences with genome tiling arrays. Science. 2004;306(5705):2242–6.
27. Zhang Y, Chen K, Sloan SA, Bennett ML, Scholze AR, O'Keeffe S, et al. An RNA-sequencing transcriptome and splicing database of glia, neurons, and vascular cells of the cerebral cortex. J Neurosci Off J Soc Neurosci. 2014;34(36):11929–47.
28. Morris KV, Mattick JS. The rise of regulatory RNA. Nat Rev Genet. 2014;15(6):423–37.
29. Sun J, Lin Y, Wu J. Long non-coding RNA expression profiling of mouse testis during postnatal development. PLoS ONE. 2013;8(10):e75750.
30. Martens-Uzunova ES, Bottcher R, Croce CM, Jenster G, Visakorpi T, Calin GA. Long noncoding RNA in prostate, bladder, and kidney cancer. Eur Urol. 2014;65(6):1140–51.
31. Du Z, Fei T, Verhaak RG, Su Z, Zhang Y, Brown M, et al. Integrative genomic analyses reveal clinically relevant long noncoding RNAs in human cancer. Nat Struct Mol Biol. 2013;20(7):908–13.
32. Beaulieu YB, Kleinman CL, Landry-Voyer AM, Majewski J, Bachand F. Polyadenylation-dependent control of long noncoding RNA expression by the poly(A)-binding protein nuclear 1. PLoS Genet. 2012;8(11):e1003078.
33. Wu Z, Liu X, Liu L, Deng H, Zhang J, Xu Q, et al. Regulation of lncRNA expression. Cell Mol Biol Lett. 2014;19(4):561–75.
34. Geisler S, Lojek L, Khalil AM, Baker KE, Coller J. Decapping of long noncoding RNAs regulates inducible genes. Mol Cell. 2012;45(3):279–91.
35. Mikkelsen TS, Ku M, Jaffe DB, Issac B, Lieberman E, Giannoukos G, et al. Genome-wide maps of chromatin state in pluripotent and lineage-committed cells. Nature. 2007;448(7153):553–60.
36. Guttman M, Amit I, Garber M, French C, Lin MF, Feldser D, et al. Chromatin signature reveals over a thousand highly conserved large non-coding RNAs in mammals. Nature. 2009;458(7235):223–7.
37. Guttman M, Rinn JL. Modular regulatory principles of large non-coding RNAs. Nature. 2012;482(7385):339–46.
38. Cawley S, Bekiranov S, Ng HH, Kapranov P, Sekinger EA, Kampa D, et al. Unbiased mapping of transcription factor binding sites along human chromosomes 21 and 22 points to widespread regulation of noncoding RNAs. Cell. 2004;116(4):499–509.
39. Sheik Mohamed MJ, Gaughwin PM, Lim B, Robson P, Lipovich L. Conserved long noncoding RNAs transcriptionally regulated by Oct4 and Nanog modulate pluripotency in mouse embryonic stem cells. RNA. 2010;16(2):324–37.
40. Dinger ME, Amaral PP, Mercer TR, Pang KC, Bruce SJ, Gardiner BB, et al. Long noncoding RNAs in mouse embryonic stem cell pluripotency and differentiation. Genome Res. 2008;18(9):1433–45.
41. Wan G, Hu X, Liu Y, Han C, Sood AK, Calin GA, et al. A novel non-coding RNA lncRNA-JADE connects DNA damage signalling to histone H4 acetylation. EMBO J. 2013;32(21):2833–47.

42. Huarte M, Guttman M, Feldser D, Garber M, Koziol MJ, Kenzelmann-Broz D, et al. A large intergenic noncoding RNA induced by p53 mediates global gene repression in the p53 response. Cell. 2010;142(3):409–19.
43. Braconi C, Kogure T, Valeri N, Huang N, Nuovo G, Costinean S, et al. microRNA-29 can regulate expression of the long non-coding RNA gene MEG3 in hepatocellular cancer. Oncogene. 2011;30(47):4750–6.
44. Benetatos L, Hatzimichael E, Dasoula A, Dranitsaris G, Tsiara S, Syrrou M, et al. CpG methylation analysis of the MEG3 and SNRPN imprinted genes in acute myeloid leukemia and myelodysplastic syndromes. Leuk Res. 2010;34(2):148–53.
45. Amort T, Souliere MF, Wille A, Jia XY, Fiegl H, Worle H, et al. Long non-coding RNAs as targets for cytosine methylation. RNA Biol. 2013;10(6):1003–8.
46. Lai F, Shiekhattar R. Where long noncoding RNAs meet DNA methylation. Cell Res. 2014;24(3):263–4.
47. Yang F, Huo XS, Yuan SX, Zhang L, Zhou WP, Wang F, et al. Repression of the long noncoding RNA-LET by histone deacetylase 3 contributes to hypoxia-mediated metastasis. Mol Cell. 2013;49(6):1083–96.
48. Yang H, Zhong Y, Xie H, Lai X, Xu M, Nie Y, et al. Induction of the liver cancer-down-regulated long noncoding RNA uc002mbe.2 mediates trichostatin-induced apoptosis of liver cancer cells. Biochem Pharmacol. 2013;85(12):1761–9.
49. Wu SC, Kallin EM, Zhang Y. Role of H3K27 methylation in the regulation of lncRNA expression. Cell Res. 2010;20(10):1109–16.
50. Zhao J, Sun BK, Erwin JA, Song JJ, Lee JT. Polycomb proteins targeted by a short repeat RNA to the mouse X chromosome. Science. 2008;322(5902):750–6.
51. Zheng GX, Do BT, Webster DE, Khavari PA, Chang HY. Dicer-microRNA-Myc circuit promotes transcription of hundreds of long noncoding RNAs. Nat Struct Mol Biol. 2014;21 (7):585–90.
52. Leucci E, Patella F, Waage J, Holmstrom K, Lindow M, Porse B, et al. microRNA-9 targets the long non-coding RNA MALAT1 for degradation in the nucleus. Sci Rep. 2013;3:2535.
53. Yoon JH, Abdelmohsen K, Srikantan S, Yang X, Martindale JL, De S, et al. LincRNA-p21 suppresses target mRNA translation. Mol Cell. 2012;47(4):648–55.
54. Chiyomaru T, Fukuhara S, Saini S, Majid S, Deng G, Shahryari V, et al. Long non-coding RNA HOTAIR is targeted and regulated by miR-141 in human cancer cells. J Biol Chem. 2014;289(18):12550–65.
55. Struhl K. Transcriptional noise and the fidelity of initiation by RNA polymerase II. Nat Struct Mol Biol. 2007;14(2):103–5.
56. van Bakel H, Nislow C, Blencowe BJ, Hughes TR. Most "dark matter" transcripts are associated with known genes. PLoS Biol. 2010;8(5):e1000371.
57. Zhu JG, Shen YH, Liu HL, Liu M, Shen YQ, Kong XQ, et al. Long noncoding RNAs expression profile of the developing mouse heart. J Cell Biochem. 2014;115(5):910–8.
58. Amaral PP, Mattick JS. Noncoding RNA in development. Mamm Genome. 2008;19(7–8):454–92.
59. Guttman M, Donaghey J, Carey BW, Garber M, Grenier JK, Munson G, et al. lincRNAs act in the circuitry controlling pluripotency and differentiation. Nature. 2011;477(7364):295–300.
60. Loewer S, Cabili MN, Guttman M, Loh YH, Thomas K, Park IH, et al. Large intergenic non-coding RNA-RoR modulates reprogramming of human induced pluripotent stem cells. Nat Genet. 2010;42(12):1113–7.
61. Mercer TR, Dinger ME, Mattick JS. Long non-coding RNAs: insights into functions. Nat Rev Genet. 2009;10(3):155–9.
62. Ponjavic J, Ponting CP, Lunter G. Functionality or transcriptional noise? Evidence for selection within long noncoding RNAs. Genome Res. 2007;17(5):556–65.
63. Ulitsky I, Shkumatava A, Jan CH, Sive H, Bartel DP. Conserved function of lincRNAs in vertebrate embryonic development despite rapid sequence evolution. Cell. 2011;147 (7):1537–50.

64. Marques AC, Ponting CP. Catalogues of mammalian long noncoding RNAs: modest conservation and incompleteness. Genome Biol. 2009;10(11):R124.
65. Ashe HL, Monks J, Wijgerde M, Fraser P, Proudfoot NJ. Intergenic transcription and transinduction of the human beta-globin locus. Genes Dev. 1997;11(19):2494–509.
66. Guenther MG, Levine SS, Boyer LA, Jaenisch R, Young RA. A chromatin landmark and transcription initiation at most promoters in human cells. Cell. 2007;130(1):77–88.
67. Wang X, Arai S, Song X, Reichart D, Du K, Pascual G, et al. Induced ncRNAs allosterically modify RNA-binding proteins in cis to inhibit transcription. Nature. 2008;454(7200):126–30.
68. Feng J, Bi C, Clark BS, Mady R, Shah P, Kohtz JD. The Evf-2 noncoding RNA is transcribed from the Dlx-5/6 ultraconserved region and functions as a Dlx-2 transcriptional coactivator. Genes Dev. 2006;20(11):1470–84.
69. Kino T, Hurt DE, Ichijo T, Nader N, Chrousos GP. Noncoding RNA gas5 is a growth arrest- and starvation-associated repressor of the glucocorticoid receptor. Sci Signal. 2010;3(107):ra8.
70. Espinoza CA, Allen TA, Hieb AR, Kugel JF, Goodrich JA. B2 RNA binds directly to RNA polymerase II to repress transcript synthesis. Nat Struct Mol Biol. 2004;11(9):822–9.
71. Espinoza CA, Goodrich JA, Kugel JF. Characterization of the structure, function, and mechanism of B2 RNA, an ncRNA repressor of RNA polymerase II transcription. RNA. 2007;13(4):583–96.
72. Mattick JS. The genetic signatures of noncoding RNAs. PLoS Genet. 2009;5(4):e1000459.
73. Khalil AM, Guttman M, Huarte M, Garber M, Raj A, Rivea Morales D, et al. Many human large intergenic noncoding RNAs associate with chromatin-modifying complexes and affect gene expression. Proc Natl Acad Sci USA. 2009;106(28):11667–72.
74. Penny GD, Kay GF, Sheardown SA, Rastan S, Brockdorff N. Requirement for Xist in X chromosome inactivation. Nature. 1996;379(6561):131–7.
75. Rinn JL, Kertesz M, Wang JK, Squazzo SL, Xu X, Brugmann SA, et al. Functional demarcation of active and silent chromatin domains in human HOX loci by noncoding RNAs. Cell. 2007;129(7):1311–23.
76. Tsai MC, Manor O, Wan Y, Mosammaparast N, Wang JK, Lan F, et al. Long noncoding RNA as modular scaffold of histone modification complexes. Science. 2010;329(5992):689–93.
77. Lyle R, Watanabe D, te Vruchte D, Lerchner W, Smrzka OW, Wutz A, et al. The imprinted antisense RNA at the Igf2r locus overlaps but does not imprint Mas1. Nat Genet. 2000;25(1):19–21.
78. Nagano T, Mitchell JA, Sanz LA, Pauler FM, Ferguson-Smith AC, Feil R, et al. The Air noncoding RNA epigenetically silences transcription by targeting G9a to chromatin. Science. 2008;322(5908):1717–20.
79. Korostowski L, Sedlak N, Engel N. The Kcnq1ot1 long non-coding RNA affects chromatin conformation and expression of Kcnq1, but does not regulate its imprinting in the developing heart. PLoS Genet. 2012;8(9):e1002956.
80. Mohammad F, Mondal T, Guseva N, Pandey GK, Kanduri C. Kcnq1ot1 noncoding RNA mediates transcriptional gene silencing by interacting with Dnmt1. Development. 2010;137(15):2493–9.
81. Mohammad F, Pandey RR, Nagano T, Chakalova L, Mondal T, Fraser P, et al. Kcnq1ot1/Lit1 noncoding RNA mediates transcriptional silencing by targeting to the perinucleolar region. Mol Cell Biol. 2008;28(11):3713–28.
82. Pandey RR, Mondal T, Mohammad F, Enroth S, Redrup L, Komorowski J, et al. Kcnq1ot1 antisense noncoding RNA mediates lineage-specific transcriptional silencing through chromatin-level regulation. Mol Cell. 2008;32(2):232–46.
83. Mohammad F, Pandey GK, Mondal T, Enroth S, Redrup L, Gyllensten U, et al. Long noncoding RNA-mediated maintenance of DNA methylation and transcriptional gene silencing. Development. 2012;139(15):2792–803.
84. Wang Y, Xu Z, Jiang J, Xu C, Kang J, Xiao L, et al. Endogenous miRNA sponge lincRNA-RoR regulates Oct4, Nanog, and Sox2 in human embryonic stem cell self-renewal. Dev Cell. 2013;25(1):69–80.

85. Willingham AT, Orth AP, Batalov S, Peters EC, Wen BG, Aza-Blanc P, et al. A strategy for probing the function of noncoding RNAs finds a repressor of NFAT. Science. 2005;309 (5740):1570–3.
86. Gregg C, Zhang J, Weissbourd B, Luo S, Schroth GP, Haig D, et al. High-resolution analysis of parent-of-origin allelic expression in the mouse brain. Science. 2010;329(5992):643–8.
87. Ayoub AE, Oh S, Xie Y, Leng J, Cotney J, Dominguez MH, et al. Transcriptional programs in transient embryonic zones of the cerebral cortex defined by high-resolution mRNA sequencing. Proc Natl Acad Sci USA. 2011;108(36):14950–5.
88. Aprea J, Prenninger S, Dori M, Ghosh T, Monasor LS, Wessendorf E, et al. Transcriptome sequencing during mouse brain development identifies long non-coding RNAs functionally involved in neurogenic commitment. EMBO J. 2013;32(24):3145–60.
89. Lv J, Cui W, Liu H, He H, Xiu Y, Guo J, et al. Identification and characterization of long non-coding RNAs related to mouse embryonic brain development from available transcriptomic data. PLoS ONE. 2013;8(8):e71152.
90. Mercer TR, Qureshi IA, Gokhan S, Dinger ME, Li G, Mattick JS, et al. Long noncoding RNAs in neuronal-glial fate specification and oligodendrocyte lineage maturation. BMC Neurosci. 2010;11:14.
91. Johnson R, Teh CH, Jia H, Vanisri RR, Pandey T, Lu ZH, et al. Regulation of neural macroRNAs by the transcriptional repressor REST. RNA. 2009;15(1):85–96.
92. Bouchard M, Grote D, Craven SE, Sun Q, Steinlein P, Busslinger M. Identification of Pax2-regulated genes by expression profiling of the mid-hindbrain organizer region. Development. 2005;132(11):2633–43.
93. Ng SY, Bogu GK, Soh BS, Stanton LW. The long noncoding RNA RMST interacts with SOX2 to regulate neurogenesis. Mol Cell. 2013;51(3):349–59.
94. Schuurmans C, Armant O, Nieto M, Stenman JM, Britz O, Klenin N, et al. Sequential phases of cortical specification involve Neurogenin-dependent and-independent pathways. EMBO J. 2004;23(14):2892–902.
95. Onoguchi M, Hirabayashi Y, Koseki H, Gotoh Y. A noncoding RNA regulates the neurogenin1 gene locus during mouse neocortical development. Proc Natl Acad Sci USA. 2012;109(42):16939–44.
96. Clarke DL, Johansson CB, Wilbertz J, Veress B, Nilsson E, Karlstrom H, et al. Generalized potential of adult neural stem cells. Science. 2000;288(5471):1660–3.
97. Lui JH, Hansen DV, Kriegstein AR. Development and evolution of the human neocortex. Cell. 2011;146(1):18–36.
98. Kornblum HI. Introduction to neural stem cells. Stroke. 2007;38(2 Suppl):810–6.
99. Ugoya SO, Tu J. Bench to bedside of neural stem cell in traumatic brain injury. Stem Cells Int. 2012;2012:141624.
100. Miljan EA, Sinden JD. Stem cell treatment of ischemic brain injury. Curr Opin Mol Ther. 2009;11(4):394–403.
101. Ramos AD, Diaz A, Nellore A, Delgado RN, Park KY, Gonzales-Roybal G, et al. Integration of genome-wide approaches identifies lncRNAs of adult neural stem cells and their progeny in vivo. Cell Stem Cell. 2013;12(5):616–28.
102. Qi Y, Cai J, Wu Y, Wu R, Lee J, Fu H, et al. Control of oligodendrocyte differentiation by the Nkx2.2 homeodomain transcription factor. Development. 2001;128(14):2723–33.
103. Tochitani S, Hayashizaki Y. Nkx2.2 antisense RNA overexpression enhanced oligodendrocytic differentiation. Biochem Biophys Res Commun. 2008;372(4):691–6.
104. Bond AM, Vangompel MJ, Sametsky EA, Clark MF, Savage JC, Disterhoft JF, et al. Balanced gene regulation by an embryonic brain ncRNA is critical for adult hippocampal GABA circuitry. Nat Neurosci. 2009;12(8):1020–7.
105. Shahryari A, Rafiee MR, Fouani Y, Oliae NA, Samaei NM, Shafiee M, et al. Two novel splice variants of SOX2OT, SOX2OT-S1, and SOX2OT-S2 are coupregulated with SOX2 and OCT4 in esophageal squamous cell carcinoma. Stem Cells. 2014;32(1):126–34.

106. Amaral PP, Neyt C, Wilkins SJ, Askarian-Amiri ME, Sunkin SM, Perkins AC, et al. Complex architecture and regulated expression of the Sox2ot locus during vertebrate development. RNA. 2009;15(11):2013–27.
107. Tsuiji H, Yoshimoto R, Hasegawa Y, Furuno M, Yoshida M, Nakagawa S. Competition between a noncoding exon and introns: Gomafu contains tandem UACUAAC repeats and associates with splicing factor-1. Genes Cells. 2011;16(5):479–90.
108. Young TL, Matsuda T, Cepko CL. The noncoding RNA taurine upregulated gene 1 is required for differentiation of the murine retina. Curr Biol. 2005;15(6):501–12.
109. Zhu CC, Dyer MA, Uchikawa M, Kondoh H, Lagutin OV, Oliver G. Six3-mediated auto repression and eye development requires its interaction with members of the Groucho-related family of co-repressors. Development. 2002;129(12):2835–49.
110. Rapicavoli NA, Poth EM, Zhu H, Blackshaw S. The long noncoding RNA Six3OS acts in trans to regulate retinal development by modulating Six3 activity. Neural Dev. 2011;6:32.
111. Lin N, Chang KY, Li Z, Gates K, Rana ZA, Dang J, et al. An evolutionarily conserved long noncoding RNA TUNA controls pluripotency and neural lineage commitment. Mol Cell. 2014;53(6):1005–19.
112. Ramos AD, Andersen RE, Liu SJ, Nowakowski TJ, Hong SJ, Gertz CC, et al. The long noncoding RNA pnky regulates neuronal differentiation of embryonic and postnatal neural stem cells. Cell Stem Cell. 2015;16(4):439–47.
113. Vance KW, Sansom SN, Lee S, Chalei V, Kong L, Cooper SE, et al. The long non-coding RNA Paupar regulates the expression of both local and distal genes. EMBO J. 2014;33 (4):296–311.
114. Chalei V, Sansom SN, Kong L, Lee S, Montiel JF, Vance KW, et al. The long non-coding RNA Dali is an epigenetic regulator of neural differentiation. Elife. 2014;3:e04530.
115. Anderson SA, Eisenstat DD, Shi L, Rubenstein JL. Interneuron migration from basal forebrain to neocortex: dependence on Dlx genes. Science. 1997;278(5337):474–6.
116. Anderson SA, Qiu M, Bulfone A, Eisenstat DD, Meneses J, Pedersen R, et al. Mutations of the homeobox genes Dlx-1 and Dlx-2 disrupt the striatal subventricular zone and differentiation of late born striatal neurons. Neuron. 1997;19(1):27–37.
117. Cobos I, Calcagnotto ME, Vilaythong AJ, Thwin MT, Noebels JL, Baraban SC, et al. Mice lacking Dlx1 show subtype-specific loss of interneurons, reduced inhibition and epilepsy. Nat Neurosci. 2005;8(8):1059–68.
118. Wang Y, Dye CA, Sohal V, Long JE, Estrada RC, Roztocil T, et al. Dlx5 and Dlx6 regulate the development of parvalbumin-expressing cortical interneurons. J Neurosci. 2010;30 (15):5334–45.
119. Zerucha T, Stuhmer T, Hatch G, Park BK, Long Q, Yu G, et al. A highly conserved enhancer in the Dlx5/Dlx6 intergenic region is the site of cross-regulatory interactions between Dlx genes in the embryonic forebrain. J Neurosci. 2000;20(2):709–21.
120. Cavallaro M, Mariani J, Lancini C, Latorre E, Caccia R, Gullo F, et al. Impaired generation of mature neurons by neural stem cells from hypomorphic Sox2 mutants. Development. 2008;135(3):541–57.
121. Favaro R, Valotta M, Ferri AL, Latorre E, Mariani J, Giachino C, et al. Hippocampal development and neural stem cell maintenance require Sox2-dependent regulation of Shh. Nat Neurosci. 2009;12(10):1248–56.
122. Askarian-Amiri ME, Seyfoddin V, Smart CE, Wang J, Kim JE, Hansji H, et al. Emerging role of long non-coding RNA SOX2OT in SOX2 regulation in breast cancer. PLoS ONE. 2014;9(7):e102140.
123. Uhde CW, Vives J, Jaeger I, Li M. Rmst is a novel marker for the mouse ventral mesencephalic floor plate and the anterior dorsal midline cells. PLoS ONE. 2010;5(1):e8641.
124. Yang Z, Ming XF. mTOR signalling: the molecular interface connecting metabolic stress, aging and cardiovascular diseases. Obes Rev. 2012;13(Suppl 2):58–68.
125. Huang Z, Wu Q, Guryanova OA, Cheng L, Shou W, Rich JN, et al. Deubiquitylase HAUSP stabilizes REST and promotes maintenance of neural progenitor cells. Nat Cell Biol. 2011;13 (2):142–52.

126. Ballas N, Grunseich C, Lu DD, Speh JC, Mandel G. REST and its corepressors mediate plasticity of neuronal gene chromatin throughout neurogenesis. Cell. 2005;121(4):645–57.
127. Rapicavoli NA, Poth EM, Blackshaw S. The long noncoding RNA RNCR2 directs mouse retinal cell specification. BMC Dev Biol. 2010;10:49.
128. Sone M, Hayashi T, Tarui H, Agata K, Takeichi M, Nakagawa S. The mRNA-like noncoding RNA Gomafu constitutes a novel nuclear domain in a subset of neurons. J Cell Sci. 2007;120 (Pt 15):2498–506.
129. Angelov D, Bondarenko VA, Almagro S, Menoni H, Mongelard F, Hans F, et al. Nucleolin is a histone chaperone with FACT-like activity and assists remodeling of nucleosomes. EMBO J. 2006;25(8):1669–79.
130. Dempsey LA, Sun H, Hanakahi LA, Maizels N. G4 DNA binding by LR1 and its subunits, nucleolin and hnRNP D, A role for G-G pairing in immunoglobulin switch recombination. J Biol Chem. 1999;274(2):1066–71.
131. Yano M, Okano HJ, Okano H. Involvement of Hu and heterogeneous nuclear ribonucleoprotein K in neuronal differentiation through p21 mRNA post-transcriptional regulation. J Biol Chem. 2005;280(13):12690–9.
132. Zheng S, Gray EE, Chawla G, Porse BT, O'Dell TJ, Black DL. PSD-95 is post-transcriptionally repressed during early neural development by PTBP1 and PTBP2. Nat Neurosci. 2012;15(3):381–8 (S1).
133. Keppetipola N, Sharma S, Li Q, Black DL. Neuronal regulation of pre-mRNA splicing by polypyrimidine tract binding proteins, PTBP1 and PTBP2. Crit Rev Biochem Mol Biol. 2012;47(4):360–78.
134. Takayama K, Horie-Inoue K, Katayama S, Suzuki T, Tsutsumi S, Ikeda K, et al. Androgen-responsive long noncoding RNA CTBP1-AS promotes prostate cancer. EMBO J. 2013;32(12):1665–80.
135. Sansom SN, Griffiths DS, Faedo A, Kleinjan DJ, Ruan Y, Smith J, et al. The level of the transcription factor Pax6 is essential for controlling the balance between neural stem cell self-renewal and neurogenesis. PLoS Genet. 2009;5(6):e1000511.

Author Biographies

Dr. Xiaomin Dong obtained his PhD degree in the Department of Gene Therapy, Ulm University, Germany in 2012, with his thesis about the application of viral vectors in the research of neurodegenerative disorders. In 2013, he joined Prof. Jiaqian Wu's laboratory in the Vivian L. Smith Department of Neurosurgery at the University of Texas as a postdoctoral fellow. His main research interests focus on the function role of long noncoding RNAs in central nervous system. He used next-generation sequencing technology to analyze the cell-type-specific long noncoding RNAs in the mouse brain. He further identified the potential functions of long noncoding RNAs during the differentiation of neural stem cells.

Dr. Naveen Reddy Muppani obtained his Master of Science degree in biotechnology from Osmania University, India, and he received his PhD in medical science from the Karolinska Insittutet. Dr. Muppani conducted his postdoctoral studies at Wright State University in Ohio and the University of Texas Health Science Center in Houston. His main research interest was focused on understanding the involvement of transcriptionally regulated cell signaling in cell survival or death implicated in human diseases/disorders such as cancer, Parkinson's disease or Alzheimer's disease. Dr. Muppani has been awarded various travel grants and fellowships (*e.g.,* Travel grant, DKFZ-KI PhD workshop on frontiers in cancer research, Germany; Travel grant, Retreat of the Helmholtz international graduate school for cancer research, Germany; and FOU Praktikant, Karolinska Institutet, Sweden). He also published articles in peer-reviewed journals and presented his research at several national and international conferences.

Prof. Jiaqian Wu An assistant professor in the Vivian L. Smith Department of Neurosurgery and Center for Stem Cell and Regenerative Medicine at the University of Texas Medical School at Houston, Dr. Wu earned her doctorate in molecular and human genetics at Baylor College of Medicine and did her postdoctoral work at Yale and Stanford University. The Wu laboratory combines stem cell biology and systems-based approaches involving functional genomics, bioinformatics, and next-generation sequencing technologies to unravel gene transcription and regulatory mechanisms governing neural and blood development and differentiation. Dr. Wu's work has been recognized with prestige honors and awards, including the National Institute of Health Pathway to Independence (PI) Award (K99/R00), R01 and the Senator Lloyd & B.A. Bentsen Investigator Award which she currently holds; the National Institutes of Health Ruth L. Kirschstein National Research Service Award for Individual Postdoctoral Fellows; and the International Society for Stem Cell Research (ISSCR) Annual Meeting Travel Award. A reviewer for NIH, MRC, the journals Nucleic Acids Research, Genome Research, and Genome Biology, Dr. Wu has presented invited talks and lectures on stem cell biology, functional genomics, and proteomics at international conferences, the Multiple Sclerosis Research Center of New York, Lawrence Livermore National Laboratory, and the University of Florida, etc. She has developed a patent, authored a book, and wrote many articles that have appeared in PNAS, Genome Biology, Plos Genetics, Genome Research, the Journal of Neuroscience, and Nature, among others.

Chapter 4
Gene Expression Models of Signaling Pathways

Jeffrey T. Chang

Abstract Aberrant pathway activation is a hallmark of a range of diseases, from atherosclerosis to diabetes to cancer. The ability to easily measure the compendium of activated pathways in a biological sample would greatly impact the study of these diseases. To do so, methods have been developed recently that leverage the gene expression profile of a cell. While these profiles provide a quantitative measure of the expression levels of every gene in the genome, they are also a reflection and amalgamation of all the processes, many of which require the concerted activity of a group of genes, that are activated in a biological sample. To interpret these profiles, methods have been developed over the last decade that can quantify the activation of individual processes. In short, each process is modeled as a gene expression signature, which is comprised of a set of genes whose expression levels are indicative of activation of the process. To score the activation of that process in a cell, computational algorithms have been developed that can compare the signature against the gene expression profile of the cell. This chapter describes the development of signatures and signature databases, as well as computational approaches to predict pathway activation in gene expression profiles.

Keywords Bioinformatics · Microarrays · Cell signaling

J.T. Chang (✉)
Department of Integrative Biology and Pharmacology, School of Medicine, Houston, USA
e-mail: jeffrey.t.chang@uth.tmc.edu

J.T. Chang
Institute of Molecular Medicine (Adjunct), School of Medicine, Houston, USA

J.T. Chang
School of Biomedical Informatics (Adjunct), University of Texas Health Science Center at Houston, Houston, TX, USA

J. Wu, *Transcriptomics and Gene Regulation*,
Translational Bioinformatics 9, DOI 10.1007/978-94-017-7450-5_4

4.1 Introduction

The development of DNA microarrays in 1995 provided the widespread ability for scientists to measure the expression level of genes across the genome [1]. The early DNA microarrays were based on glass slides, in which oligonucleotide probes targeting each transcript were affixed. The cDNA from samples of interest could be hybridized to the probes. By measuring a detectable marker, such as a fluorescent molecule, that is conjugated to the cDNA, a quantitative measure that correlates with the amount of each transcript in the sample could be detected. Since then, additional technologies have been developed, such as procedures creating probes using a photolithography method (such as currently used in Affymetrix microarrays) [2–5], or using bead-based methods (used in Illumina microarrays) [6]. Although microarrays have been successful, they are currently being supplanted for the detection of gene expression by technologies using next-generation sequencing such as RNA-Seq [7]. Nevertheless, microarrays are still being used due to their cost advantage. Regardless, although the technologies to measure gene expression have progressed rapidly over the last 20 years, the end product is the same: a relative measure of the gene expression values across a genomic scale. In addition, even though the technology to measure the gene expression levels differs, the broad principles for how to interpret these values remain the same.

The interpretation of gene expression profiles has evolved in parallel with the technologies. With the ability to measure the gene expression level of all genes on a genome scale came along the desire to compare the genes across different conditions. That is, a screen for gene expression changes by looking for differential gene expression across biological or experimentally derived conditions. Early examples of the types of experiments involving differential gene expression are the identification of genes that changed conditions between normal and disease states [8], cycles in the cell cycle [9, 10], and other biological studies [11, 12].

4.2 Machine Learning

Although the study of the gene expression changes of individual genes is important, it was quickly realized that gene expression profiles contain additional information. That is, the patterns of gene expression changes across a set of genes are also informative. The biological justification for this is that biological processes, such as cell proliferation, motility, immune response, are carried forth not by the activity of a single gene, but by the multiple genes acting in concert. This led to the deduction that the analysis of patterns of gene expression, of multiple genes working together, can provide insight into the working of a cell.

To find patterns of gene expression, *clustering* methods were developed. Clustering methods can find patterns that are apparent when looking at a group of data samples [13, 14]. That is, a single sample can provide the relative expression of genes in that sample. With two samples, the changes in expression of a gene can be determined (although in practice, many replicates of the samples are done to reduce noise). However, with multiple samples, one can compare the patterns of expression (i.e., in which samples the expression is up or down) across a set of genes. This is known as a guilt-by-association inference [15–18] (Fig. 4.1). The interpretation of this inference is that if two genes show a similar pattern of expression across a set of samples (i.e., they are activated and have high expression in the same samples, and are also repressed with low expression in the other samples), then they are likely to be regulated by the same mechanism. Given these conditions, the genes are associated transcriptionally, and we infer that they may be involved in the same process. Thus, if we know the function of one gene, then we can use that to predict the function of the other. This approach is commonly used to predict the function of unannotated genes, although there are limitations to this method [19].

To find genes that have similar gene expression profiles, clustering methods are used. At the same time, the algorithm can also detect groups of samples that have same profiles. Not only are the genes then predicted to have the same functions, but also some similarity can be inferred for the samples, because similar groups of genes are activated. This has been used to great effect in cancer genomics. One of the first successes of this approach is the clustering of breast cancers into different subtypes [20]. Using an unbiased measure based on transcriptional profiles, breast cancer was split into four subtypes that exhibit different levels of activation. These subtypes have subsequently been refined [21]. Nevertheless, the different subtypes of breast cancers were found to exhibit very distinct and clinically relevant behaviors. The HER2+ and basal subtypes lead to the worst outcomes [22], and recommendations are being crafted to vary patient treatments depending on the subtype of their disease [23]. This demonstrates that the patterns of gene expression profiles can be applied in an unbiased manner to learn about biology and guide clinical treatment.

Reflecting the fact that clustering is unbiased, it is also sometimes called unsupervised machine learning, as opposed to supervised machine learning methods [24]. Both unsupervised and supervised machine learning are concerned with finding patterns of data. The difference between these two approaches is whether

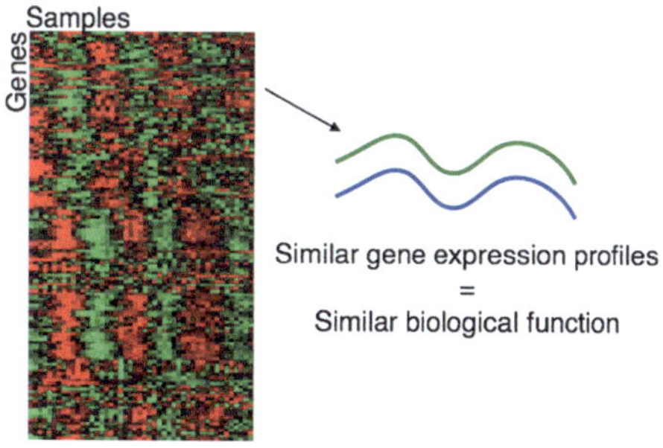

Fig. 4.1 Guilt by association

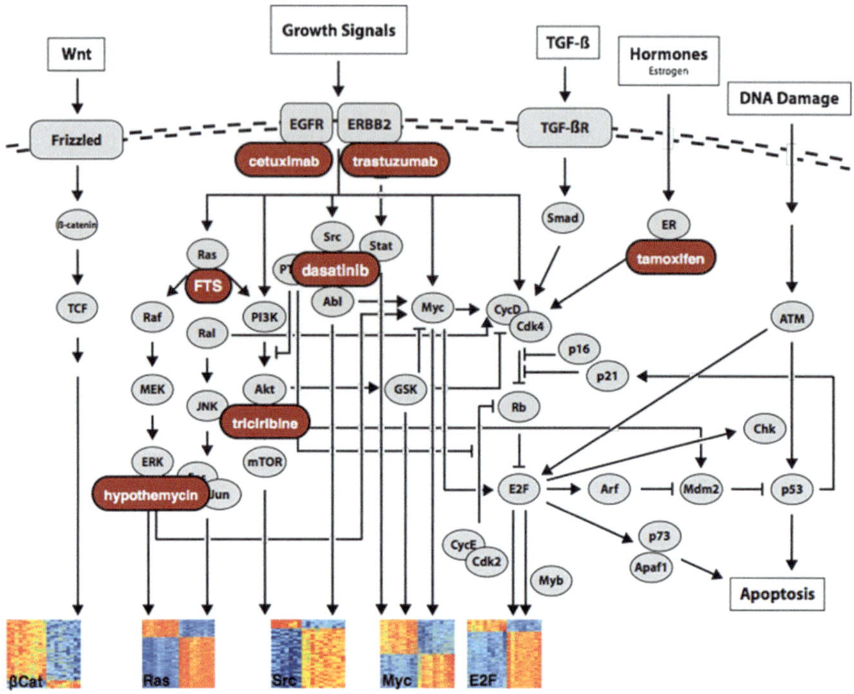

Fig. 4.2 Clinical relevance of gene expression signatures

the algorithm is searching for patterns in a completely unbiased manner (for unsupervised), or searching for the presence or absence of a previously known pattern (for supervised machine learning). In contrast to unsupervised clustering, a supervised machine learning algorithm starts with a specific expression profile of interest and then examples whether a test sample also exhibits that expression profile. The existence of supervised machine learning methods means that previously established expression profiles can be detected in other samples. That is, if one has an expression profile, consisting of genes that are somehow altered, one can measure its presence in another data set. This forms the basis of pathway analysis.

4.3 Pathway Activation in Gene Expression

After the invention of microarrays, it has been observed that essentially every biological process leaves an imprint on the gene expression profile of a cell. While it might have been possible that gene expression was limited to the measurements of activities that involved transcriptional changes, a surprising result is the ability for gene expression profiles to also reflect other post-transcriptional changes. Many biological processes are regulated on a post-transcriptional level, such as plasma

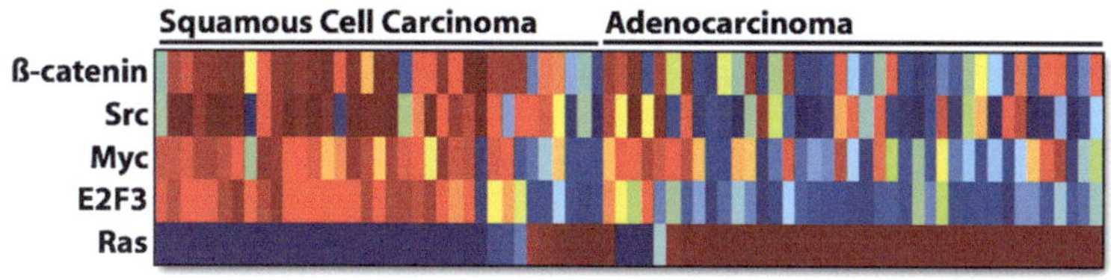

Fig. 4.3 Activation of pathways across tumor samples

membrane receptor activation, kinase cascades, and protein localization events. However, essentially all of these events eventually result in a cascade of changes that lead to changes in gene expression levels.

Among many events that lead to changes in gene expression profiles, the activation of pathways does as well [25] (Fig. 4.2). Aberrant pathway activation is a hallmark of a range of diseases, and this has probably been best seen in cancer [26, 27] (Fig. 4.3). While cancer is driven by well-known oncogenes and tumor suppressor genes, they affect pathways that can also be perturbed by mutations in many other genes. While there are thousands of genes have been reported to be mutated in cancer, the mutations affect a much more limited number of core pathways [27]. Many of the genes that are altered are poorly annotated, and therefore, it is unknown what pathway is affected by their mutations. Nevertheless, because a knowledge of the altered pathway can affect prognosis and treatment, there is a great interest in understanding which are the pathways affected by the mutant genes.

4.4 Pathway Definitions

Although the term *pathway* is commonly used, for gene expression analysis, it is helpful to separate the molecular components of the pathway into two classes. In one, a pathway consists of the sequence of molecules that form a signal transduction activity. We call the molecules that comprise this sequence as the *members* of the pathway (Fig. 4.4). These members are related by biochemical relationships, such as physical binding, phosphorylation, and ubiquitination. For example, if we consider the Rb pathway, which controls the G_1/s *check point*, as curated in BioCarta [28], we see that it includes many members. Rb drives the control of cell cycle progression into S phase and is considered a major tumor suppressor pathway. The Rb protein itself is a member of the pathway [29, 30]. Upstream of it, Rb is phosphorylated by cyclins, including Cyclin D and Cyclin E, which are also considered members. Upon phosphorylation, it derepresses the E2F transcription factors (also members), which then transcriptionally activates genes that allow for cell cycle progression.

Next, we define the second type of components of a pathway. These are molecules that are their *transcriptional effectors*. These are the genes whose expressions are increased or decreased upon activation of the pathway. Continuing the example above, the effectors of the Rb pathway would be the genes whose expression

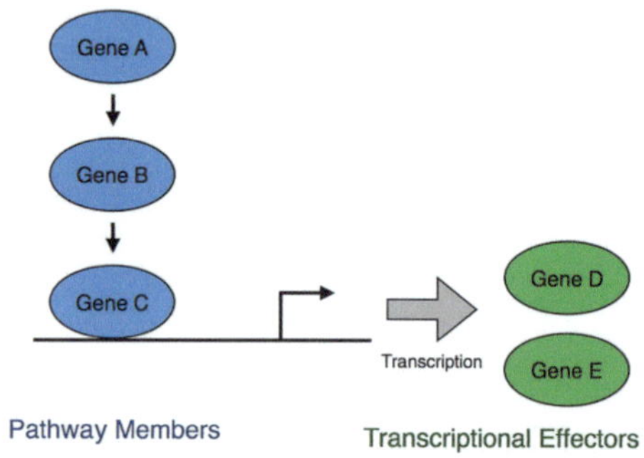

Fig. 4.4 Pathway members and effectors

changes in response to activation of Rb, specifically, the transcriptional targets of the E2F transcription factors, including E2F1 itself, DHFR, TK, or CCNE1 [31, 32]. In other words, every transcriptional target of E2Fs would comprise the Rb pathway. E2F1 is a transcriptional effector because it induces its own expression. Rb would not be, since its expression is unchanged upon its activation.

Distinguishing the members from effectors of a pathway is useful for gene expression analysis of pathways. The members of a pathway are not necessarily regulated transcriptionally. When the pathway is activated or inactivated, the expression levels of the proteins do not change. Therefore, when predicting the activation of the pathway, the expression levels of the members are not necessarily good markers (although they are often used that way in practice). However, the transcriptional effectors are viable indicators of pathway activation. Changes in the effectors can provide clues to the status of the pathway. The methods that can integrate and score their gene expression levels are described below

4.5 Pathway Databases and Representations

Pathways are comprised of transcriptional effectors that can be used in gene expression signatures. While the pathway diagrams from typical databases also show edges that indicate the types (and often the direction) of interactions among molecules, that information is typically not used in the calculation of gene expression signatures. Only the identities of the genes are important. Thus, pathways are represented by lists of genes, or sometimes called *bags of genes* or gene sets, underscoring the fact that the genes are unstructured and unordered (Fig. 4.5).

One limitation to the bag of gene representation is that it does not account for the fact that some genes may be more conclusive markers of pathway activation than others. To do this, it is possible to add a weight to the genes for a signature that reflects their importance. In this way, the relative importance of a gene to predict the activation of a pathway can be modeled. The expression levels of the genes can be combined by a computational method that can account for the weights to predict the outcomes. Sometimes, a weighted list of genes is called a gene expression signature.

Fig. 4.5 Representations of pathways

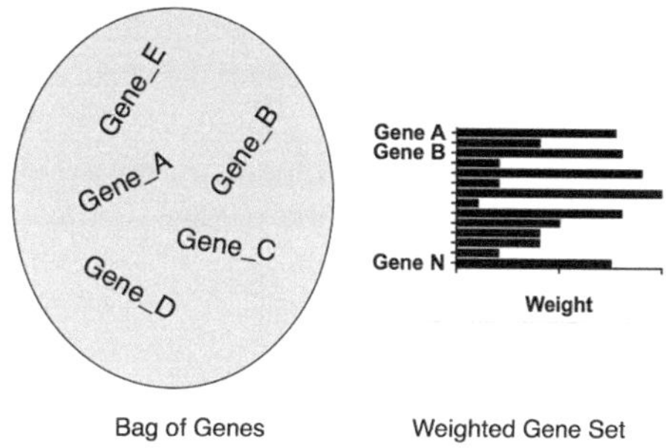

Curated lists of pathways are contained in many databases, such as KEGG [33, 34], BioCarta [28], Reactome [35], or Nature-NCI Pathway Interaction Database [36]. In these databases, curators have collected pathways from the literature (or other databases). There are also attempts, such as Pathway Commons, to provide a comprehensive set of pathways by combining them from other sources [37]. One of the limitations of these databases is the fact that the notion of a pathway is fluid and varies from source to source. There is no agreement on what constitutes a single pathway, and also no clear consensus on the boundaries between pathways. Therefore, there are differences in the pathways among the databases.

There are also databases that contain gene sets, such as GeneSigDB [38] or MSigDB (Molecular Signatures Database) [39]. MSigDB is comprised of gene sets that are curated from the literature, as well as other sources. This database does not contain only the gene sets with pathway members or effectors. Instead, it also consists of gene sets for other phenomena, such as the genes residing on specific chromosome bands and gene ontology functions [40]. However, it is currently the most comprehensive database of gene sets for pathways.

Finally, there are also databases that contain weighted gene expression signatures for pathways, such as SIGNATURE [41]. The SIGNATURE database consists of gene expression signatures for oncogenic pathways, including pathways such as Myc, Ras, EGFR, and E2F. The signatures contain specifically the transcriptional effectors, coupled with weights indicating their relative importance to predicting pathway activation. This has been used for predicting response to targeted therapies [42] and identifying subtypes of cancers [43], as well as decomposing pathways into modular structures [44].

4.6 Measuring Pathway Activation

A range of tools have been developed that can score the activation of pathways from gene expression data. One of the most straightforward approaches, and the first to be developed, is exemplified by tools such as DAVID [45] or GATHER [46]. Here, a gene expression data set covering a biological condition of interest is first processed and converted into a gene set. For example, if the goal is to compare the pathways in two different conditions, such as tissues with and without a disease, the genes that are differentially expressed in the conditions are extracted and converted into a bag of gene-type gene set. This can be done by setting an explicit cutoff, where genes with an expression change beyond this cutoff are in the gene set, while those with less change are not. In practice, sophisticated statistical tests are used [47–52]. Once a gene set representing the data set of interest is created, it is compared against the gene sets representing the transcriptional effectors of pathways. The statistical significance of the overlap can be scored using a Fisher's exact test [53], a hypergeometric distribution [54], a Bayesian test [46], or a variant of one of these [55]. The gene sets of pathways that achieve statistical significance represent the pathways that are associated with the disease.

There is a commercial pathway analysis database called Ingenuity Pathway Analysis that is frequently used. The pathway database was specifically developed by the company and human curated. A pathway analysis here results in a list of statistically significant pathways, as well as figures that show their topologies.

One of the limitations in the approach to identify pathways above is that it requires a cutoff to select the gene set that represents the biological data set. This assumes that biological processes are driven by the genes with the largest changes in gene expression. However, it is possible that processes are driven by a multitude of genes with smaller gene expression changes. Under this model of pathway activation, it would be more correct to look for the coordinated changes of a group of genes, even if their gene expression changes are not very big. To do this, enrichment algorithms, such as gene set enrichment analysis (GSEA), have been developed [56–59]. To perform a GSEA analysis, the biological condition of interest must be represented by a gene expression data set with exactly two conditions, such as the disease and control samples described above. Given this data set, GSEA can apply a gene set and determine whether the genes in the gene set *overall* exhibit a change. Specifically, GSEA scores the association of a gene set with a gene expression data set using an adapted Kolmogorov–Smirnov test. By analyzing all the gene sets from a database of pathway gene sets (typically those from the MSigDB), GSEA can find pathways that are differentially regulated across two conditions.

There are two limitations of GSEA. First, it scores whether a pathway is differentially activated across two conditions. It does not have the capability to score the activation of a pathway in individual samples. Second, it treats each gene in the gene set as being equally important and does not consider that some genes may have a stronger ability to predict pathway activation than others.

To address the first limitation of GSEA, methods have been developed to apply GSEA to single samples, such as ssGSEA [60] or GSVA [61]. These are variants of GSEA where the significance of each pathway is based on the absolute expression levels in each sample, rather than the difference between two groups of samples. This allows the calculation of a score for individual samples. The calculation of the statistical significance is done in a similar way to the standard GSEA algorithm.

To address both limitations of GSEA, one can also use quantitative gene expression signatures [62–71], like those in the SIGNATURE system described above [41–43]. This system uses a supervised machine learning framework to score the activation of pathways. In such systems, the gene expression profile that exemplifies an activated pathway is generated from a *training set* that consists of gene expression profiles of conditions in which pathways are on or off (Fig. 4.6). The pathway off state is typically cells in quiescence, while the pathway on state is cells in which a single pathway is activated by ectopic activation of the pathway. Using this training set, machine learning algorithms can make quantitative models of the activation of the pathway. The SIGNATURE system is based on logistic regression [72, 73], although many other algorithms, such as singular value decomposition-based regressions, have been used [74]. Using supervised machine learning algorithms, it is possible to score the pathway in a new sample. The main limitation of this system is the need for a quantitative gene set, created from a training set. This requires specially created training sets. Because of this extra requirement, database of gene sets, such as MSigDB, cannot be used. Therefore, there are many fewer profiles available. The largest study with these pathways uses up to 52 profiles [75].

In addition to gene expression information, other types of genomic information can also provide hints as to whether a pathway is activated. Due to the rise of technologies that can easily collect other types of genomic information, such as copy number alterations or mutations, efforts have been developed to combine all of this information to predict the activation of a pathway. An early project to do this is

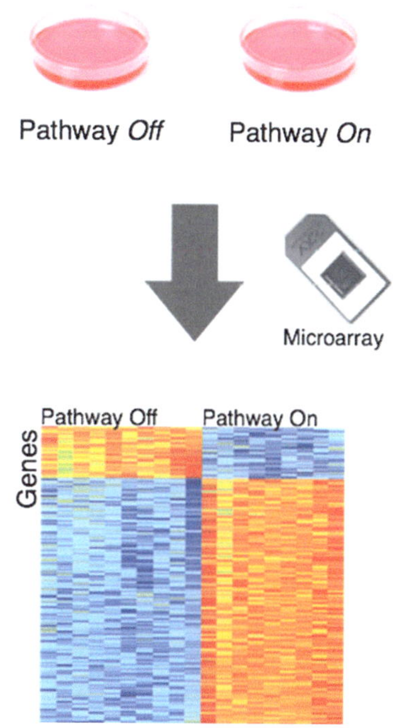

Fig. 4.6 Gene expression signatures

called PARADIGM [76]. Here, a Bayesian network is used to model pathways using the structures provided by the NCI PID database. Then, given genomic data that indicate the level of gene expression or copy number variation, the model can predict the level of activation of the pathway. This has been shown to identify patterns across groups of tumors that can classify them into distinct survival categories.

4.7 Conclusions

There is a range of representations and approaches for analyzing the activation of pathways in gene expression signatures. The theoretical foundation underlying these approaches is the idea that the gene expression profile of a cell is a reflection of the underlying activities occurring in the cell. That is, while many cellular activities happen on the post-transcriptional level and do not immediately involve transcriptional regulation, eventually the signaling pathways lead to changes in the transcription of genes that can be detected in gene expression profiles.

There are some areas in the gene expression analysis of pathways which still require further development. First, there are clear differences in how pathways work across tissues and cell types. This fact is well understood in cancer cells. For example, EGFR inhibitors have been seen to be effective in colorectal cancers [77–79]. However, they have not seen the same success in breast cancer, even in patients with over-expression of EGFR [80–82]. Therefore, there is clearly a qualitative difference in how the pathway works across these two tissues. Trying to understand the differences in pathways across cell types, and thus the differences in their gene expression signatures, might affect the prediction of pathway activation from gene expression signatures. This is still an under-explored area.

Another important question in the analysis of gene expression signatures is the degree of variation that is seen across the signatures. That is, there are many different sets of genes that can predict the activation of the same biological process. Three independent signatures that can all predict the prognosis of breast cancer data sets have very few genes in common [83]. The signatures are robust in that they can predict prognosis across data sets, so it is not that their biology is completely different. Furthermore, many distinct gene sets can be created that predict prognosis. This shows that the ability to predict a biological function with a gene expression profile can be robust, even while the identities of the genes in the signature vary.

Nevertheless, despite their limitations, gene expression signatures are starting to find their way into clinical use, particularly in cancer treatments. Oncotype DX is a gene expression signature, clinically implemented as an RT–PCR assay, that is used to predict the outcomes in node-negative cancers [84]. Also in breast cancer, subtypes can be determined with a PAM50 test [22]. This demonstrates the power of gene expression signatures to assay the underlying biology of a sample.

The ability of patterns of genes to be able to score biological phenotypes and predict clinical outcomes is an unanticipated consequence of the genomic assays developed to measure the expression of genes. Because of the development of computational technologies to interpret the gene expression changes, gene expression signatures of the activation of pathways and other biological processes have become a portable currency that can be used to compare the biological state of different samples. In other words, gene expression signatures are markers that can be used to interrogate the function of a cell, including the activation of pathways.

References

1. Schena M, Shalon D, Davis RW, Brown PO. Quantitative monitoring of gene expression patterns with a complementary DNA microarray. Science. 1995;270:467–70.
2. Fodor SP, Read JL, Pirrung MC, Stryer L, Lu AT, et al. Light-directed, spatially addressable parallel chemical synthesis. Science. 1991;251:767–73.
3. Lipshutz RJ, Fodor SP, Gingeras TR, Lockhart DJ. High density synthetic oligonucleotide arrays. Nat Genet. 1999;21:20–4.
4. Lipshutz RJ, Morris D, Chee M, Hubbell E, Kozal MJ, et al. Using oligonucleotide probe arrays to access genetic diversity. Biotechniques. 1995;19:442–7.
5. Lockhart DJ, Dong H, Byrne MC, Follettie MT, Gallo MV, et al. Expression monitoring by hybridization to high-density oligonucleotide arrays. Nat Biotechnol. 1996;14:1675–80.
6. Gunderson KL, Kruglyak S, Graige MS, Garcia F, Kermani BG, et al. Decoding randomly ordered DNA arrays. Genome Res. 2004;14:870–7.
7. Wang Z, Gerstein M, Snyder M. RNA-Seq: a revolutionary tool for transcriptomics. Nat Rev Genet. 2009;10:57–63.
8. Perou CM, Jeffrey SS, van de Rijn M, Rees CA, Eisen MB, et al. Distinctive gene expression patterns in human mammary epithelial cells and breast cancers. Proc Natl Acad Sci U S A. 1999;96:9212–7.
9. Spellman PT, Sherlock G, Zhang MQ, Iyer VR, Anders K, et al. Comprehensive identification of cell cycle-regulated genes of the yeast Saccharomyces cerevisiae by microarray hybridization. Mol Biol Cell. 1998;9:3273–97.
10. Whitfield ML, Sherlock G, Saldanha AJ, Murray JI, Ball CA, et al. Identification of genes periodically expressed in the human cell cycle and their expression in tumors. Mol Biol Cell. 2002;13:1977–2000.
11. Hughes TR, Marton MJ, Jones AR, Roberts CJ, Stoughton R, et al. Functional discovery via a compendium of expression profiles. Cell. 2000;102:109–26.
12. Lamb J, Crawford ED, Peck D, Modell JW, Blat IC, et al. The Connectivity Map: using gene-expression signatures to connect small molecules, genes, and disease. Science. 2006;313:1929–35.
13. de Hoon MJ, Imoto S, Nolan J, Miyano S. Open source clustering software. Bioinformatics. 2004;20:1453–4.
14. Eisen MB, Spellman PT, Brown PO, Botstein D. Cluster analysis and display of genome-wide expression patterns. Proc Natl Acad Sci U S A. 1998;95:14863–8.
15. Lee HK, Hsu AK, Sajdak J, Qin J, Pavlidis P. Coexpression analysis of human genes across many microarray data sets. Genome Res. 2004;14:1085–94.
16. Quackenbush J. Genomics. Microarrays–guilt by association. Science. 2003;302:240–1.
17. Staudt LM, Brown PO. Genomic views of the immune system*. Annu Rev Immunol. 2000;18:829–59.

18. Stuart JM, Segal E, Koller D, Kim SK. A gene-coexpression network for global discovery of conserved genetic modules. Science. 2003;302:249–55.
19. Gillis J, Pavlidis P. "Guilt by association" is the exception rather than the rule in gene networks. PLoS Comput Biol. 2012;8:e1002444.
20. Perou CM, Sorlie T, Eisen MB, van de Rijn M, Jeffrey SS, et al. Molecular portraits of human breast tumours. Nature. 2000;406:747–52.
21. Herschkowitz JI, Simin K, Weigman VJ, Mikaelian I, Usary J, et al. Identification of conserved gene expression features between murine mammary carcinoma models and human breast tumors. Genome Biol. 2007;8:R76.
22. Parker JS, Mullins M, Cheang MC, Leung S, Voduc D, et al. Supervised risk predictor of breast cancer based on intrinsic subtypes. J Clin Oncol. 2009;27:1160–7.
23. Goldhirsch A, Wood WC, Coates AS, Gelber RD, Thurlimann B, et al. Strategies for subtypes–dealing with the diversity of breast cancer: highlights of the St. Gallen International Expert Consensus on the Primary Therapy of Early Breast Cancer 2011. Ann Oncol. 2011;22:1736–47.
24. Bishop C. Pattern Recognition and Machine Learning. New York: Springer-Verlag; 2007.
25. Huang E, Ishida S, Pittman J, Dressman H, Bild A, et al. Gene expression phenotypic models that predict the activity of oncogenic pathways. Nat Genet. 2003;34:226–30.
26. Ding L, Getz G, Wheeler DA, Mardis ER, McLellan MD, et al. Somatic mutations affect key pathways in lung adenocarcinoma. Nature. 2008;455:1069–75.
27. Jones S, Zhang X, Parsons DW, Lin JC, Leary RJ, et al. Core signaling pathways in human pancreatic cancers revealed by global genomic analyses. Science. 2008;321:1801–6.
28. Nishimura D. BioCarta. Biotech Softw Internet Rep. 2001;2:117–120.
29. Helin K. Regulation of cell proliferation by the E2F transcription factors. Curr Opin Genet Dev. 1998;8:28–35.
30. Sherr CJ. Cancer cell cycles. Science. 1996;274:1672–7.
31. DeGregori J, Kowalik T, Nevins JR. Cellular targets for activation by the E2F1 transcription factor include DNA synthesis- and G1/S-regulatory genes. Mol Cell Biol. 1995;15:4215–24.
32. Ohtani K, DeGregori J, Leone G, Herendeen DR, Kelly TJ, et al. Expression of the HsOrc1 gene, a human ORC1 homolog, is regulated by cell proliferation via the E2F transcription factor. Mol Cell Biol. 1996;16:6977–84.
33. Kanehisa M, Goto S. KEGG: kyoto encyclopedia of genes and genomes. Nucleic Acids Res. 2000;28:27–30.
34. Tanabe M, Kanehisa M. Using the KEGG database resource. Curr Protoc Bioinformatics Chapter. 2012;1(Unit1):12.
35. Croft D, Mundo AF, Haw R, Milacic M, Weiser J, et al. The Reactome pathway knowledgebase. Nucleic Acids Res. 2014;42:D472–7.
36. Schaefer CF, Anthony K, Krupa S, Buchoff J, Day M, et al. PID: the pathway interaction database. Nucleic Acids Res. 2009;37:D674–9.
37. Cerami EG, Gross BE, Demir E, Rodchenkov I, Babur O, et al. Pathway Commons, a web resource for biological pathway data. Nucleic Acids Res. 2011;39:D685–90.
38. Culhane AC, Schwarzl T, Sultana R, Picard KC, Picard SC, et al. GeneSigDB–a curated database of gene expression signatures. Nucleic Acids Res. 2010;38:D716–25.
39. Liberzon A, Subramanian A, Pinchback R, Thorvaldsdottir H, Tamayo P, et al. Molecular signatures database (MSigDB) 3.0. Bioinformatics. 2011;27:1739–40.
40. Ashburner M, Ball CA, Blake JA, Botstein D, Butler H, et al. Gene ontology: tool for the unification of biology. The Gene Ontology Consortium. Nat Genet. 2000;25:25–9.
41. Chang JT, Gatza ML, Lucas JE, Barry WT, Vaughn P, et al. SIGNATURE: a workbench for gene expression signature analysis. BMC Bioinformatics. 2011;12:443.
42. Bild AH, Yao G, Chang JT, Wang Q, Potti A, et al. Oncogenic pathway signatures in human cancers as a guide to targeted therapies. Nature. 2006;439:353–7.
43. Gatza ML, Lucas JE, Barry WT, Kim JW, Wang Q, et al. A pathway-based classification of human breast cancer. Proc Natl Acad Sci U S A. 2010;107:6994–9.

44. Chang JT, Carvalho C, Mori S, Bild AH, Gatza ML, et al. A genomic strategy to elucidate modules of oncogenic pathway signaling networks. Mol Cell. 2009;34:104–14.
45. da Huang W, Sherman BT, Lempicki RA. Systematic and integrative analysis of large gene lists using DAVID bioinformatics resources. Nat Protoc. 2009;4:44–57.
46. Chang JT, Nevins JR. GATHER: a systems approach to interpreting genomic signatures. Bioinformatics. 2006;22:2926–33.
47. Anders S, Huber W. Differential expression analysis for sequence count data. Genome Biol. 2010;11:R106.
48. Efron B, Tibshirani R, Storey JD, Tusher V. Empirical Bayes Analysis of a Microarray Experiment. J Am Stat Assoc. 2001;96:1151–60.
49. Love MI, Huber W, Anders S. Moderated estimation of fold change and dispersion for RNA-seq data with DESeq2. Genome Biol. 2014;15:550.
50. McCarthy DJ, Chen Y, Smyth GK. Differential expression analysis of multifactor RNA-Seq experiments with respect to biological variation. Nucleic Acids Res. 2012;40:4288–97.
51. Robinson MD, Smyth GK. Moderated statistical tests for assessing differences in tag abundance. Bioinformatics. 2007;23:2881–7.
52. Trapnell C, Roberts A, Goff L, Pertea G, Kim D, et al. Differential gene and transcript expression analysis of RNA-seq experiments with TopHat and Cufflinks. Nat Protoc. 2012;7:562–78.
53. Fleiss JL. Statistical methods for rates and proportions. New York: John Wiley; 1981.
54. Fury W, Batliwalla F, Gregersen PK, Li W. Overlapping probabilities of top ranking gene lists, hypergeometric distribution, and stringency of gene selection criterion. Conf Proc IEEE Eng Med Biol Soc. 2006;1:5531–4.
55. Hosack DA, Dennis G Jr, Sherman BT, Lane HC, Lempicki RA. Identifying biological themes within lists of genes with EASE. Genome Biol. 2003;4:R70.
56. Mootha VK, Lindgren CM, Eriksson KF, Subramanian A, Sihag S, et al. PGC-1alpha-responsive genes involved in oxidative phosphorylation are coordinately downregulated in human diabetes. Nat Genet. 2003;34:267–73.
57. Subramanian A, Kuehn H, Gould J, Tamayo P, Mesirov JP. GSEA-P: a desktop application for Gene Set Enrichment Analysis. Bioinformatics. 2007;23:3251–3.
58. Subramanian A, Tamayo P, Mootha VK, Mukherjee S, Ebert BL, et al. Gene set enrichment analysis: a knowledge-based approach for interpreting genome-wide expression profiles. Proc Natl Acad Sci U S A. 2005;102:15545–50.
59. Wu D, Smyth GK. Camera: a competitive gene set test accounting for inter-gene correlation. Nucleic Acids Res. 2012;40:e133.
60. Verhaak RG, Tamayo P, Yang JY, Hubbard D, Zhang H, et al. Prognostically relevant gene signatures of high grade serous ovarian carcinoma. J Clin Invest. 2013;123:517–25.
61. Hanzelmann S, Castelo R, Guinney J. GSVA: gene set variation analysis for microarray and RNA-seq data. BMC Bioinformatics. 2013;14:7.
62. Barbie DA, Tamayo P, Boehm JS, Kim SY, Moody SE, et al. Systematic RNA interference reveals that oncogenic KRAS-driven cancers require TBK1. Nature. 2009;462:108–12.
63. Furge KA, Chen J, Koeman J, Swiatek P, Dykema K, et al. Detection of DNA copy number changes and oncogenic signaling abnormalities from gene expression data reveals MYC activation in high-grade papillary renal cell carcinoma. Cancer Res. 2007;67:3171–6.
64. Huang F, Reeves K, Han X, Fairchild C, Platero S, et al. Identification of candidate molecular markers predicting sensitivity in solid tumors to dasatinib: rationale for patient selection. Cancer Res. 2007;67:2226–38.
65. Langenau DM, Keefe MD, Storer NY, Guyon JR, Kutok JL, et al. Effects of RAS on the genesis of embryonal rhabdomyosarcoma. Genes Dev. 2007;21:1382–95.
66. Loboda A, Nebozhyn M, Klinghoffer R, Frazier J, Chastain M, et al. A gene expression signature of RAS pathway dependence predicts response to PI3 K and RAS pathway inhibitors and expands the population of RAS pathway activated tumors. BMC Med Genomics. 2010;3:26.

67. Ooi CH, Ivanova T, Wu J, Lee M, Tan IB, et al. Oncogenic pathway combinations predict clinical prognosis in gastric cancer. PLoS Genet. 2009;5:e1000676.
68. Rhodes DR, Kalyana-Sundaram S, Tomlins SA, Mahavisno V, Kasper N, et al. Molecular concepts analysis links tumors, pathways, mechanisms, and drugs. Neoplasia. 2007;9:443–54.
69. Singh A, Greninger P, Rhodes D, Koopman L, Violette S, et al. A gene expression signature associated with "K-Ras addiction" reveals regulators of EMT and tumor cell survival. Cancer Cell. 2009;15:489–500.
70. Wong DJ, Liu H, Ridky TW, Cassarino D, Segal E, et al. Module map of stem cell genes guides creation of epithelial cancer stem cells. Cell Stem Cell. 2008;2:333–44.
71. Zhang XH, Wang Q, Gerald W, Hudis CA, Norton L, et al. Latent bone metastasis in breast cancer tied to Src-dependent survival signals. Cancer Cell. 2009;16:67–78.
72. Spang R, Zuzan H, West M, Nevins J, Blanchette C, et al. Prediction and uncertainty in the analysis of gene expression profiles. In Silico Biol. 2002;2:369–81.
73. West M, Blanchette C, Dressman H, Huang E, Ishida S, et al. Predicting the clinical status of human breast cancer by using gene expression profiles. Proc Natl Acad Sci U S A. 2001;98:11462–7.
74. Liu Z, Wang M, Alvarez JV, Bonney ME, Chen CC, et al. Singular value decomposition-based regression identifies activation of endogenous signaling pathways in vivo. Genome Biol. 2008;9:R180.
75. Gatza ML, Silva GO, Parker JS, Fan C, Perou CM. An integrated genomics approach identifies drivers of proliferation in luminal-subtype human breast cancer. Nat Genet. 2014;46:1051–9.
76. Vaske CJ, Benz SC, Sanborn JZ, Earl D, Szeto C, et al. Inference of patient-specific pathway activities from multi-dimensional cancer genomics data using PARADIGM. Bioinformatics. 2010;26:i237–45.
77. Chung KY, Shia J, Kemeny NE, Shah M, Schwartz GK, et al. Cetuximab shows activity in colorectal cancer patients with tumors that do not express the epidermal growth factor receptor by immunohistochemistry. J Clin Oncol. 2005;23:1803–10.
78. Cunningham D, Humblet Y, Siena S, Khayat D, Bleiberg H, et al. Cetuximab monotherapy and cetuximab plus irinotecan in irinotecan-refractory metastatic colorectal cancer. N Engl J Med. 2004;351:337–45.
79. Saltz LB, Meropol NJ, Loehrer PJ Sr, Needle MN, Kopit J, et al. Phase II trial of cetuximab in patients with refractory colorectal cancer that expresses the epidermal growth factor receptor. J Clin Oncol. 2004;22:1201–8.
80. Baselga J, Albanell J, Ruiz A, Lluch A, Gascon P, et al. Phase II and tumor pharmacodynamic study of gefitinib in patients with advanced breast cancer. J Clin Oncol. 2005;23:5323–33.
81. Carey LA, Rugo HS, Marcom PK, Mayer EL, Esteva FJ, et al. TBCRC 001: randomized phase II study of cetuximab in combination with carboplatin in stage IV triple-negative breast cancer. J Clin Oncol. 2012;30:2615–23.
82. von Minckwitz G, Jonat W, Fasching P, du Bois A, Kleeberg U, et al. A multicentre phase II study on gefitinib in taxane- and anthracycline-pretreated metastatic breast cancer. Breast Cancer Res Treat. 2005;89:165–72.
83. Ein-Dor L, Kela I, Getz G, Givol D, Domany E. Outcome signature genes in breast cancer: is there a unique set? Bioinformatics. 2005;21:171–8.
84. Paik S, Shak S, Tang G, Kim C, Baker J, et al. A multigene assay to predict recurrence of tamoxifen-treated, node-negative breast cancer. N Engl J Med. 2004;351:2817–26.

Author Biography

Dr. Jeffrey T. Chang is an assistant professor of Integrative Biology and Pharmacology in the School of Medicine, Institute of Molecular Medicine, and School of Biomedical Informatics in the University of Texas Health Science Center at Houston (UTHSC) in Houston, TX. He received his PhD in Biomedical Informatics at Stanford University studying under Dr. Russ Altman. From there, he did his postdoctoral studies in cancer genomics with Dr. Joseph Nevins at Duke University. His research focuses on cell signaling networks in cancer, with a special emphasis on breast cancer and metastasis. Dr. Chang has published 49 scientific papers with an h-index of 27 and written five book chapters.

Chapter 5
From Gene Expression to Disease Phenotypes: Network-Based Approaches to Study Complex Human Diseases

Quanwei Zhang, Wen Zhang, Rubén Nogales-Cadenas, Jhin-Rong Lin, Ying Cai and Zhengdong D. Zhang

Abstract Gene expression is a fundamental biological process under tight regulation at all levels in normal cells. Its dysregulation can cause abnormal cell behaviors and result in diseases, and thus gene expression profiling and analysis have been widely used to provide the first clue about the molecular mechanisms of human diseases. Because genes and their products interact with and regulate one another, it is essential to analyze gene expression data and understand the genetics of disease in a biological network context. In this chapter, we first introduce the state-of-the-art gene expression analysis (GEA) with network integration and the joint analysis of mRNA and miRNA expression to understand disease regulatory mechanisms and then discuss how disease genes are predicted by incorporating knowledge of gene regulation and characterized in biological networks.

Keywords Gene expression · Disease phenotypes · Biological networks

5.1 Introduction

In the central dogma of biology, gene expression is the intermediate, critical step at which genetic information flows from DNA to functional gene products such as proteins and noncoding RNAs through RNA transcription and translation in each cell. It is the key step where various types of gene regulation—including DNA modification, transcriptional regulation, and posttranscriptional modification—take place. Gene regulation receives and spreads signals in the form of gene regulatory networks (GRNs), in which a group of genes interact with each other and control certain cell functions. Dysregulated gene expression in the network due to promoter

---

Q. Zhang · W. Zhang · R. Nogales-Cadenas · J.-R. Lin · Y. Cai · Z.D. Zhang (✉)
Department of Genetics, Albert Einstein College of Medicine, Michael F. Price Center for Genetic and Translational Medicine, 1301 Morris Park Avenue, Bronx, NY 10461, USA
e-mail: zhengdong.zhang@einstien.yu.edu

J. Wu, *Transcriptomics and Gene Regulation*,
Translational Bioinformatics 9, DOI 10.1007/978-94-017-7450-5_5

mis-methylation [1, 2], changed transcription factor levels [3, 4], mutated transcriptional regulatory elements (TREs) [5], and miRNA deregulations [6] can result in abnormal cell behaviors and have all been observed in human diseases. Considered as intermediate phenotypes, mRNA expression profiles have been analyzed in biological networks to identify causal genes of human diseases in many studies [7, 8]. In particular, among gene products, microRNAs (miRNAs) are small noncoding RNAs overrepresented in GRNs [9, 10]. Recent studies have revealed their striking gene regulatory activities at the posttranscriptional level [11] and their profound involvement in human diseases [12].

The prevailing assumption about human diseases is that the disease phenotypes are the outcome of interactions between genes and environment [13]. Linking disease phenotypes to genotypes is thus fundamental to understanding human diseases. Linkage analysis has been effective to study disorders with Mendelian inheritance patterns. To date, over 3000 genes with mutations linked to disease phenotypes are cataloged in the Online Mendelian Inheritance in Man (OMIM) database [14]. However, in contrast to Mendelian diseases with simple genetic architectures, complex diseases are characterized by the multifactorial nature and epistasis, in which the causal effects of many risk genes are obscure and cannot be effectively detected by traditional approaches [15, 16]. Furthermore, unlike Mendelian disorders where mutations usually occur within protein coding regions, the majority of mutations of complex diseases occur in noncoding regions associated with gene expression regulation [17, 18]. Deciphering the relationship between genotypes and phenotypes for complex diseases thus requires incorporating the knowledge of gene expression regulation.

Over the last decade, the Encyclopedia of DNA Elements (ENCODE) Consortium has been exploring the functional elements in the human genome and has generated comprehensive data for gene regulation such as transcription factor binding sites and gene–locus interactions [19]. This knowledge provides important basis for analyzing genetic factors of complex diseases. On the other hand, newly developed high-throughput technologies can generate genomic data with an increasingly large sample size and will certainly improve the statistical power to detect subtle associations in complex diseases. This shift has made it possible to tackle the challenges of deciphering complex diseases. With the abundance of genomic data and knowledge of gene regulation, nevertheless, new approaches are needed to integrate genomic data and knowledge of gene regulation to connect genotypes and phenotypes of complex diseases.

Most proteins exert their functions through interactions with other proteins. Such inter- and intracellular interconnectivity implies that the impact of a specific genetic variation is not restricted to the activity of the gene product that carries it, but can spread along the links of the network and alter the activity of other related gene products that otherwise carry no changes. Therefore, an understanding of gene/protein network context is essential to understand the genetics of disease. With the advent of next-generation sequencing, the throughput and the resolution of gene expression profiling have both been increased to an unprecedented level. In addition to traditional methods of gene expression analysis (GEA), network-based approaches to GEA have also been developed [20–23]. Incorporation of network information into the estimation procedure of the regression model not only encourages smoothness in the estimate of contributions of candidate genes but also integrates into its calculation a priori biological information from the network, which is ignored in conventional methods. A network-based method for gene set enrichment analysis has been developed. Combining a graph-based statistic with an interactive sub-network visualization, EnrichNet takes into account the network structure of physical interactions between the gene sets of interest and improves the prioritization of putative gene set associations as well as exploits information from molecular interaction networks and gene expression data [24]. NetworkAnalyst, another software tool, can perform network analysis and visualization given a gene list. It can also consider multiple meta-data parameters to perform a meta-analysis of multiple gene expression datasets [25].

Not only can disease genes be identified with network-integrated methods, but also they can be studied as a whole in the context of biological networks. Most biological networks are scale-free networks whose degree distribution follows a power law: $P(X = x) = x^{-\alpha}$, in which x is the node degree and α is a constant. In a scale-free network, a small number of nodes tend to have higher degree (such nodes are called hubs), while a large number of nodes have low degrees. Generally, we can divide commonly used network characteristics into different levels. On the gene (protein) level, degree, closeness centrality, and betweenness centrality are often used. They measure, respectively, the number of its interactions, its centeredness in the network, and its importance in communication between genes. On the neighborhood level, clustering coefficient is widely used to measure the probability that the neighbors of a node are connected with one another. On the gene pairs level, one of the most used characteristics is the shortest path between two nodes. Studies of the network characteristics of a group of related disease genes can provide us insights into the molecular mechanisms of the disease.

5.2 Gene Expression Analysis with Network Integration

Gene expression analysis (GEA) has been widely used in human disease studies. High-throughput technologies to profile gene expression include DNA microarrays, serial analysis of gene expression, quantitative RT-PCT, differential-display RT-PCR, and parallel signature sequencing [26]. Network-based GEA is an efficient way to analyze gene expression data because it takes advantage of the functional relationship among genes or their products.

Networks are particularly valuable for modeling large-scale biological systems and have been used with increasing frequency to analyze such complex systems. Graph theory provides useful mathematical tools for general network analysis [27], which can be easily adapted to study genes and pathways. Here, we introduce a class of regression methods with network integration, focusing on the difference between their approaches and applications. We first introduce linear regression with network regularization. We then present a network-regularized logistic regression method. We next describe a network-regularized Cox model. And finally, we summarize the application results.

5.2.1 Linear Regression Methods with Network Regularization

One issue in GEA is the high dimensionality of the transcriptomic data, e.g., the number of covariates (genes) is much larger than that of observations (samples) [28]. Providing a straightforward mathematical framework for variation indications, linear models have been widely used in data analysis [28]. The biological network can be described as a graph by its adjacency or Laplacian matrix and provides crucial and complementary biological information to gene expression data. A novel linear regression method governed by Laplacian network-deduced matrix has been proposed to identify molecular pathways from gene expression data [20]. In this method, a network-constrained penalty function is used to penalize the L_1-norm of regression coefficients [20]. The method is in essence a mathematical programming problem whose solution criterion is $\hat{\theta} = \arg\min_\theta \mathbb{C}(\theta, \lambda, \alpha)$, in which $\hat{\theta}$ is the estimated contribution coefficient of each gene, $\mathbb{C}(\theta, \lambda, \alpha)$ is the network-constrained regularization criterion defined in [20], λ and α are the two parameters to be defined through a leave-one-out cross-validation (CV) process.

5.2.2 Network-Regularized Logistic Regression Method

For classification problems with gene expression data, Logit-Lapnet was put forward to identify molecular pathways associated with breast cancer [21]. It is a regression method combining logistic models and network regularization with the graphical Laplacian matrix. The data matrix is derived from gene expression profiles. The L_1-normed regularization and the corresponding extensions, elastic net and fused lasso, have been used to identify molecular pathways. Extending the previous similar approaches, the Logit-Lapnet method incorporates a priori functional information contained in biological networks. We can consider Logit-Lapnet in a simple way, i.e., as a logistic regression method regularized by lasso and network two items. Its model estimation is formulated as a convex optimization problem, guaranteeing the identifiability of an optimal solution (Fig. 5.1). The optimization criteria, $L(\lambda, \alpha, \beta)$, contains the generalized L_2-norm penalty term using the Laplacian graphical matrix, which encourages smoothness on contribution coefficients (see [21] for a quantitative description of the grouping effect on Logit-Lapnet concerning the structure of network).

$$L(\lambda,\alpha,\beta)=\sum_{i=1}^{n}[-y_iX_i\beta+ln(1+e^{X_i\beta})]+\lambda\alpha|\beta|_1+\lambda(1-\alpha)\beta^T L\beta$$

$$\hat{\beta}=\arg\min_{\beta}\mathrm{L}(\lambda,\alpha,\beta)$$

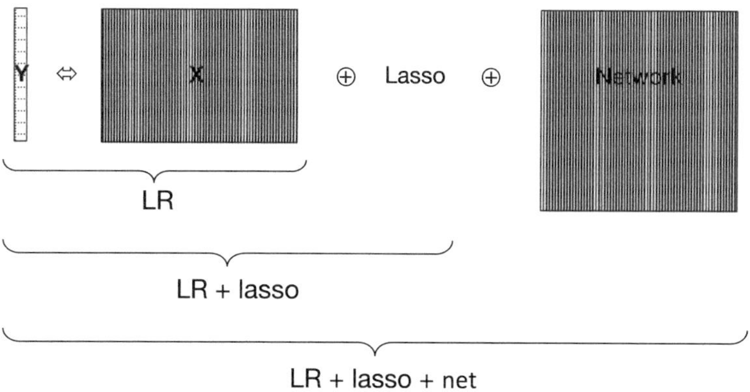

Fig. 5.1 Logit-Lapnet optimization criteria

5.2.3 Network-Regularized Cox Model and Its Application

For survival analysis of gene expression data, a Cox proportional hazard model with network regularization was used to select connected network modules predictive of survival of breast cancer patients [29]. Its optimization criterion to estimate gene contribution is a modified likelihood function of the Cox model: $h(t,x_j)=h_0(t)e^{x_j^T\beta}$, in which $h_0(t)$ is the baseline hazard function at time t, x_j the vector of biomarkers for genes, and β the gene coefficient vector. The estimation is defined as $\hat{\beta}=\arg\min_{\beta}\mathbb{C}(\lambda,\alpha,\beta)$, in which $\mathbb{C}(\lambda,\alpha,\beta)$ contains the negative log likelihood function with L_1 + L_2 norm and network regularizations on the coefficient vector. The new Cox model showed better performance in simulation than conventional Cox models and was much more sensitive to cancer-related genes and network modules. Genes identified by the new Cox model have clear biological functions involving cancer cell apoptosis and cell cycle.

5.2.4 Application Results

Performance assessment by simulation demonstrated that Logit-Lapnet outperforms elastic net and lasso, two alternative methods (Fig. 5.2) [21]. Application of network-regularized linear regression methods to glioblastoma gene expression data

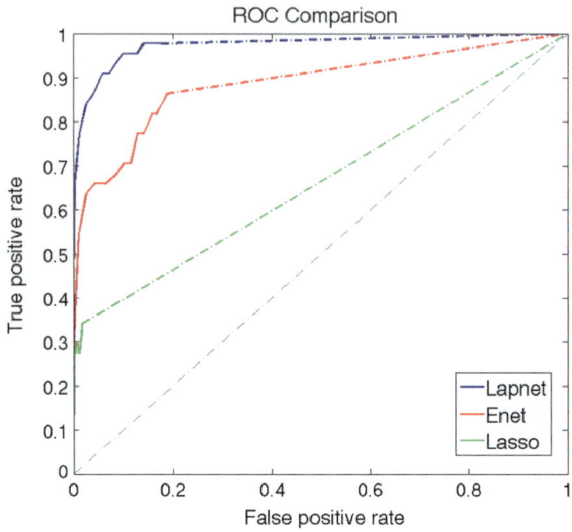

Fig. 5.2 Performance assessment by simulations

identified pathways that might be related to cancer survival time [20]. In a study of biomarkers for breast cancer, Logit-Lapnet selected 262 genes, 166 ($\sim$63 %) of which interact with one another (Fig. 5.3). By comparison, lasso selected only 24 genes, 20 of which are isolated, while elastic net selected 393 genes, 232 ($\sim$59 %) of which are interconnected [21]. The advantage of network-regularized Cox model was demonstrated by its application to breast cancer gene ascertainment [29], in which it selected more known mutated cancer biomarkers than the conventional means.

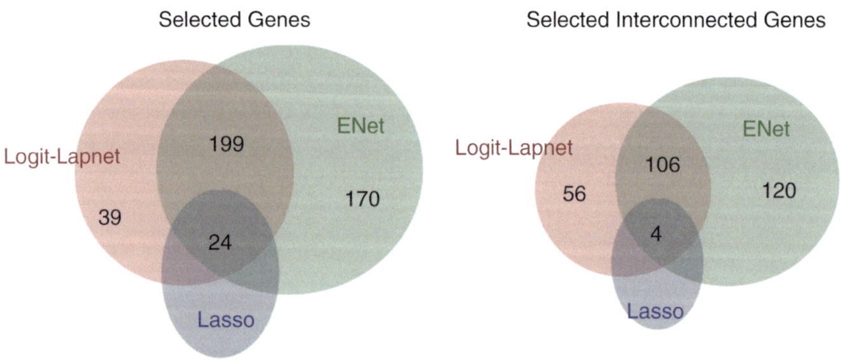

Fig. 5.3 Gene numbers selected by Logit-Lapnet, lasso, and elastic net

5.3 Analyzing Expression of mRNAs and miRNAs to Understand Disease Regulatory Mechanisms

Microarray- and sequencing-based gene expression profiling has been widely used to investigate complex diseases including cancer. Recent studies have discovered gene signatures of numerous diseases and biomarkers for prognosis prediction and disease sub-type classification. For example, Wang et al. [30] and van't Veer et al. [31], respectively, identified ~70 genes that predict breast cancer metastasis risk. Parker et al. [32] proposed a 50-gene PAM50 model, commonly used for breast cancer classification. These markers include genes that control cell cycle, proliferation, DNA replication, and repair, many of which are differentially expressed due to genomic mutations affecting transcriptional regulation.

Testing for differentially expressed genes can yield up to thousands of candidate genes, and one common way to study their functions is to analyze their enrichment in biological pathways. Because the experimentally validated canonical pathways (such as KEGG pathways) are largely incomplete [33], functional interpretation of the candidate genes based on them can be misleading. A less biased approach is based on biological networks, especially those derived from high-throughput data. It can reveal interactions among genes or gene products beyond pathways and has been shown to outperform methods for breast cancer metastasis prediction based on differential expression analysis only [34]. Co-expression networks and GRNs are two representative biological networks widely used to interpret mRNA expression data in disease phenotypes (Fig. 5.4). They are often constructed or inferred for each individual experiment and hence reveal cell type or conditional specific knowledge. In addition, many tools for network-based analysis and visualization have been developed, including GeneMANIA [35] and Cytoscape [36].

Among gene regulatory mechanisms, miRNAs have recently been revealed as one of the most important factors. miRNAs are small noncoding RNA molecules whose main function is to silence gene expression, mainly through transcription repression or mRNA degradation. They are known to be key regulators in important cellular processes such as development [37] and cycle progression [38]. In recent years, they have gained importance in different aspects of human disease research: as targets of miR mimics [39] or antagomirs [40] to reverse disease progression, as biomarkers to detect diseases [41, 42], and as drugs to improve the effect of already developed treatments [43]. Hence, mRNAs and miRNAs regulatory networks analyses are complementary, and both have become indispensable in the study of complex human diseases.

5.3.1 Co-expression Network

Co-expression networks aim at finding genes sharing similar expression patterns across diverse conditions by measuring the correlation of expression between each

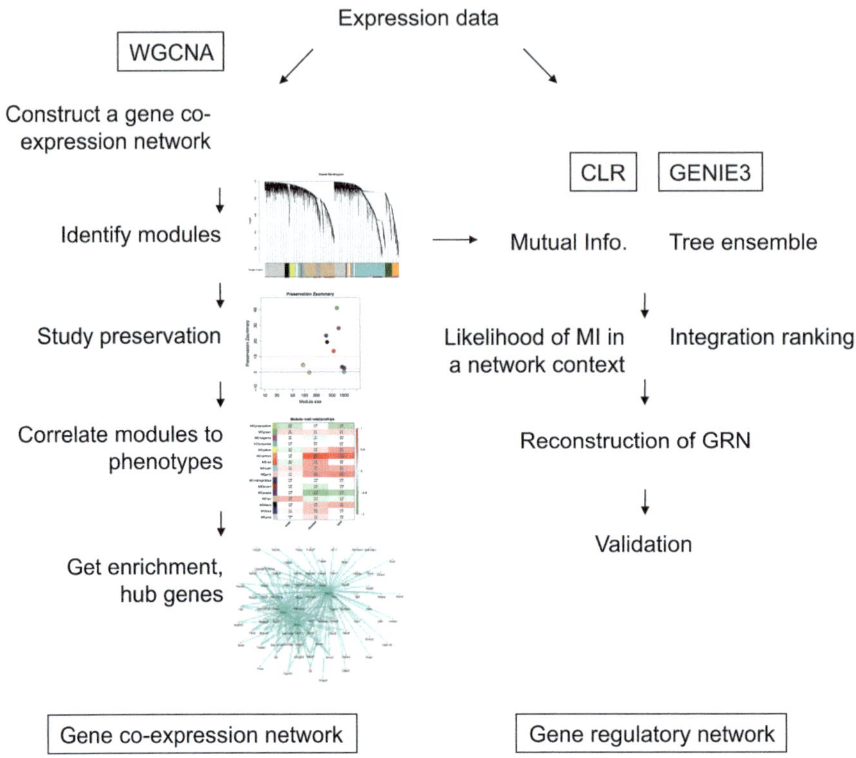

Fig. 5.4 Gene expression data analysis with gene networks

pair of genes, under the assumption that they function together in tightly connected biology processes. The weighted gene co-expression network analysis (WGCNA) [44] is now a popular way to find modules—i.e., groups of genes—as higher-order expression patterns and disease signatures. Gene–gene correlations are first quantified by Pearson's correlation coefficient, and modules are then identified using a topological overlap measure algorithm. A composite Z summary statistic indicates module preservation: whether the modules are robust in different conditions and independent datasets. One can then find contribution made by highly preserved modules to certain trait by measuring correlation coefficient between module eigengene value (the first principal component) and quantitative phenotypes. Hub genes (i.e., genes with many connections) in such modules are important. The WGCNA has been mostly used in developmental studies, where there are no controls and samples are usually arranged in a time course, such as hematopoietic stem cell ontogeny [45] and brain neuron formation [46]. Databases such as GeneMANIA [35] and COXPRESdb [47], which compile assorted datasets, are good co-expression data sources for query genes of interest.

5.3.2 Genetic Regulatory Network

Reconstruction of GRNs is an age-old challenge. Various algorithms can achieve this, but no single method shows the optimal performance across all datasets [48]. One of the well-established methods is context likelihood of relatedness (CLR), an extension of the relevance network technique based on mutual information (MI) [49]. The approach first scores the MI between each pair of a transcriptional regulator (TR) and its potential target gene, and then scores the likelihood of the regulation within its network context; those with high values are likely to form a regulatory relationship. Because a TR may regulates its targets in a nonlinear way, mutual information is a better choice than correlation for not requiring linearity or continuity of the dependence. In addition, the CLR method can be combined with WGCNA to find TRs in modules [45]. Recently, the DREAM4 in Silico network challenge [48] compared over 30 GRN-inference methods for high-throughput data. GENIE3 [50], a random forest-based method, is one of the top-performing methods. It treats GRN inference as a feature selection problem and predicts the expression of a target gene from the expression of all other genes (input genes) using random forests or extra-trees machine learning approaches. The contribution of an input gene on target gene expression is used to build the putative regulatory links. After aggregating links from all genes, the whole GRN is reconstructed from ranked interactions. Databases such as RegulonDB [51] provide experimentally confirmed regulatory interactions that can also validate the accuracy of the GRN inference methods.

5.3.3 miRNAs Regulation in Human Disease

Studies have implicated miRNAs in many diverse illnesses such as hepatitis B and C [52, 53], cardiac and heart diseases [54, 55], and even behavior and neuronal system diseases such as Tourette's syndrome [56]. In particular, important is the study of miRNAs in cancer research, as they are known to regulate important processes in cancer biology such as angiogenesis [57], apoptosis [58], and cell differentiation [59]. Here, we describe the common principle of these analyses—the integration of miRNAs and mRNAs expression, sequence pairwise information, and functional information.

miRNA regulation analysis. miRNAs regulate gene transcriptional activity by total or partial matching of nucleotide sequences with targeted mRNAs. Many computational algorithms are available to predict miRNA targets based on different criteria such as base pairing and target accessibility [60–62]. In general, their predictions are considered to be complementary and are usually combined to increase the overall sensitivity of the prediction [63, 64]. Each method, however, suffers from high false-positive and false-negative rates [65]. This happens even with the inclusion of experimental validated interactions from databases such as miRWalk [66] or miRecords [67]. Thus, the predicted mRNA–miRNA interactions should be considered as working hypotheses, since they do not necessarily fit with the disease phenotypes. In the study of disease gene regulation, it is advisable to integrate these predictions not only with differential expression values of mRNAs from case and control individuals, but also with miRNAs expression values.

Identification of miRNA regulatory mechanisms. Regulatory mechanisms of biological processes generally involve more than one miRNA and mRNA functioning together. Many computational approaches have been proposed to identify such regulatory mechanisms. They differ from one another in their methodological approaches and their usage of mRNA/miRNA expression values and external information such as potentially involved pathways. Methods used in different contexts include Bayesian networks [68], probabilistic methods [69], LASSO regression [70], or rule-based methods [71]. Despite their differences, the overall analytical flow of these methods is similar (Fig. 5.5).

Functional analysis of miRNA regulation. It is common to infer the function of a miRNA from its gene targets (for possible bias in such an approach, see [72]).

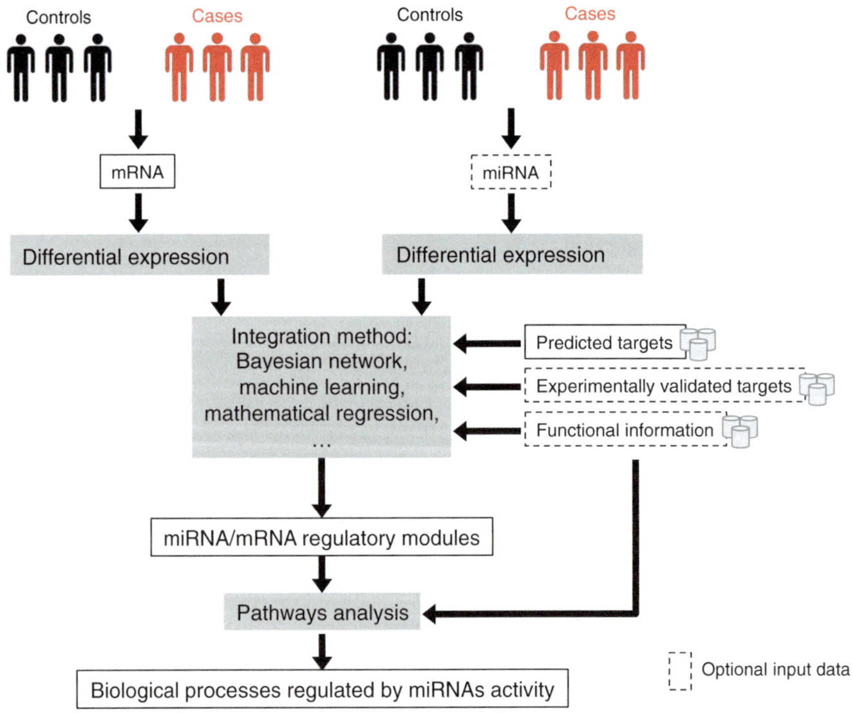

Fig. 5.5 miRNA analysis pipeline

The incorporation of external information, such as functional terms related to mRNA targets, makes it possible to deduce the involvement of miRNAs regulation in biological pathways [73]. This strategy can be used to interpret functional enrichment results and to find regulatory modules of miRNAs–mRNAs participating in the same processes [74]. Several resources provide direct functional annotation of miRNAs (Table 5.1).

Table 5.1 Useful resources of miRNA regulation for human disease studies

Resource	Description	Ref.
MiRNA database		
miRBase	Database of miRNA sequences and annotations for 206 different organisms	[99]
miRNA-target interaction		
microRNA.org	Database of predicted microRNA targets & target down-regulation scores. It includes experimentally observed expression patterns	[100]
miRWalk	Database that provides information on miRNA from human, mouse and rat on their predicted as well as validated binding sites on their target genes. It includes information on experimentally validated miRNA interaction information associated with genes, pathways, diseases, organs, OMIM disorders, cell lines, and literature on miRNAs	[66]
multiMiR	R package and database for miRNA-target interaction which includes information based on disease annotation and drug microRNA response, in addition to many experimental and computational databases	[101]
CancerMiner	Database including recurring microRNA-mRNA associations across cancer type	[102]
Functional information		
mir2Disease	A manually curated database providing a comprehensive resource of miRNA deregulation in various human diseases	[103]
mirFocus	Database providing leads for in-depth analysis of miRNA-target gene pathways and the related miRNA annotations	www mirfocus.org
HmDD	Database with curated experiment-supported evidence for human microRNA (miRNA) and disease associations	[104]
miRCancer	Database providing a collection of miRNA expression profiles in various human cancers, automatically extracted from the published literatures in PubMed	[105]
Variant information		
PolymiRTS	Database of naturally occurring DNA variations in microRNA (miRNA) seed regions and miRNA-target sites underlying in gene expression and disease phenotypes	[106]
miRdSNP	Data source of dSNPs and robust tools to capture their spacial relationship with miRNA-target sites on the 3'UTRs of human genes	[107]

5.4 Predicting Disease Genes by Incorporating Knowledge of Gene Regulation

The identification of disease genes is a fundamental objective in medical research. With the advent of high-throughput genotyping technologies, a large number of disease-associated variants have been identified by genome-wide association studies (GWASs) [75]. Such disease variants provide valuable signals for uncovering underlying disease genes and unraveling disease mechanisms, which can be improved by leveraging the knowledge of gene regulation.

5.4.1 Importance of Knowledge of Gene Regulation in Complex Disease Prediction

Both genetic predisposition and environmental factors may contribute to the pathogenesis of complex diseases. The origins of genetic predisposition are genetic variants that affect gene functions and thus contribute to disease susceptibility. Some of these variants are located in coding regions and affect gene functions by altering the corresponding protein sequences. The others, located in noncoding regions, may affect (TREs), such as transcription factor binding sites, resulting in dysregulation of gene expression.

Uncovering disease causal genes that underlie the association signals discovered in GWAS is challenging. The simplest method is to select genes closest to disease-associated variants as the causal genes. However, because single nucleotide polymorphisms (SNPs) used in GWAS are tagging SNPs, representing linkage disequilibrium (LD) blocks, disease-associated SNPs discovered in GWAS are most likely not causal SNPs but mere their proxies. Another more sophisticated method is to first define the LD regions tagged by GWAS SNPs and then identify genes overlapping LD regions as candidate causal genes [76]. Causal genes near GWAS SNPs are likely to be included in this way. However, causal genes whose expression is affected by causal SNPs through modifying their TREs will almost certainly be missed, as they fall outside LD regions. To include these "distal" causal genes, it requires knowledge of gene regulation and, more specifically, knowledge of regulatory relationship between loci and genes.

5.4.2 Gene Regulation Data Resources and Complex Disease Risk Loci

Studies have shown that disease-associated SNPs are overrepresented in loci implicated in gene regulations [77–79] (Fig. 5.6). There are several important resources for the knowledge of aforementioned gene-locus regulation linkage.

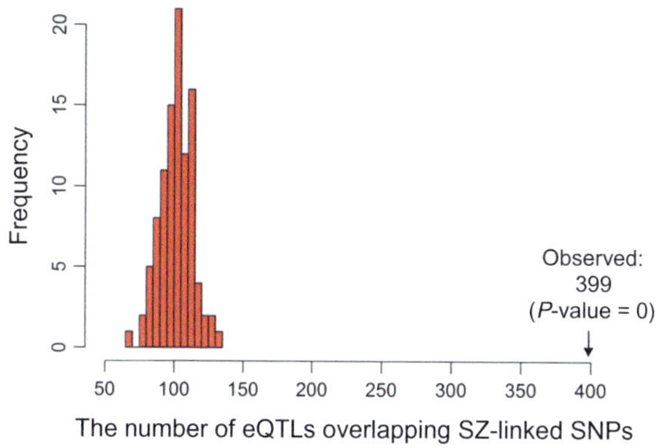

Fig. 5.6 Enrichment of schizophrenia-associated SNPs at eQTLs. We compiled 125,568 eQTLs from GTEx studies and identified 15,027 SNPs in high linkage disequilibrium with 261 schizophrenia-associated SNPs that we collected from the GWAS catalog [111] and a meta-analysis of schizophrenia [76]. 399 eQTLs are SZ-linked SNPs (P = 0, permutation test with 100 repetitions)

Expression quantitative trait loci (eQTL) are genomic loci whose genotypes are associated with transcript levels. eQTL data provide valuable information of gene-locus regulatory relationship and are useful in prioritizing GWAS signals [80]. In addition, the ENCODE Project inferred regulatory relationship from correlation between DNase I hypersensitivity of loci and promoters in different cell and tissue types [81]. Furthermore, FANTOM5 generated regulation information between enhancers and target genes by comparing their transcriptional activities across different cell types [78]. These regulatory data repositories serve as important information resources for not only prioritizing but also exploring new disease causal factors, on both SNP and gene levels.

5.4.3 *Linking Distal Candidate Causal Genes by Incorporating the Knowledge of Gene Regulation*

As mentioned earlier, causal genes may not always fall in the same haplotype block carrying GWAS SNPs, and thus, it requires other information in addition to LD to identify them. Figure 5.7 shows an example of successfully uncovering a promising causal gene underlying a GWAS SNP by using the gene regulatory information. SNP rs2159767 is a GWAS SNP associated with schizophrenia [82]. The LD region indexed by rs2159767 is in a gene desert and thus devoid of any genes. In it, however, we found two TREs that are likely to regulate two distal genes, fragile X mental retardation 1 (FMR1), and fragile X mental retardation 1 neighbor (FMR1NB), respectively. Notably, FMR1 is a literature-supported SZ gene [83],

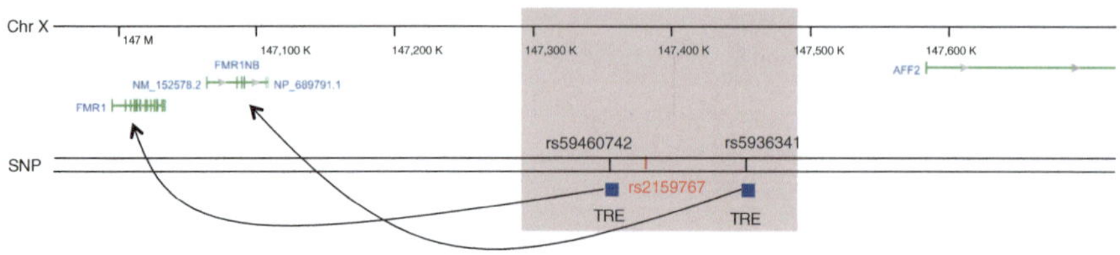

Fig. 5.7 Distal disease causal gene candidates. Gene regulatory information can link genes far away from the disease-associated GWAS SNP (schizophrenia-associated rs2159767 in this case) to the disease risk region (the *red block*)

and we found that a SNP (rs59460742) within the TRE associated with FMR1 is in strong LD ($r^2 = 0.587$) with rs2159767. Those evidences imply that the causal factor of the GWAS signal could be the SNP within the TRE that results in the dysregulation of FMR1.

5.4.4 Distal and Proximal Candidate Causal Genes

In general, incorporating LD information can improve the detection of causal genes in the proximity of GWAS signals, but finding distal causal genes relies on the knowledge of gene regulation. Using LD and gene regulation information, we identified three overlapping sets of candidate causal genes for schizophrenia (Fig. 5.8). There are 485 proximal and 158 distal candidate causal genes. Together, these two numbers indicate that incorporating gene regulatory information can substantially expand the set of candidate causal genes (about one-third in the aforementioned schizophrenia case). Although irrelevant distal genes could be

Fig. 5.8 Schizophrenia causal gene candidates. Candidates genes are linked to 261 schizophrenia-associated SNPs through different gene regulatory information

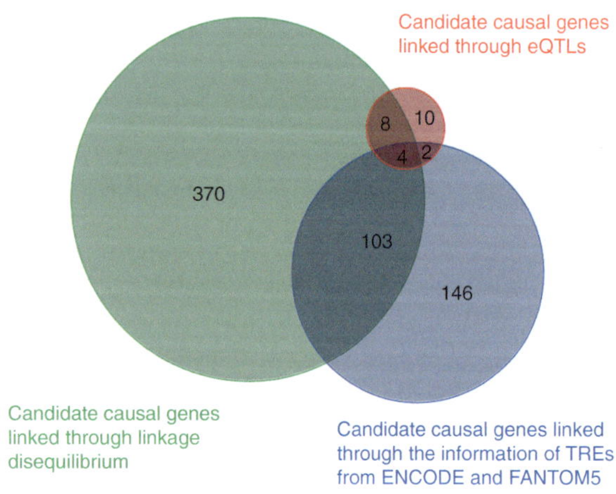

introduced due to false regulatory linkage, incorporating the knowledge of gene regulation can cover potential risk genes in a more comprehensive manner, which will also facilitate the downstream analysis.

5.5 Characterizing the Network and Association Properties of Disease Genes

Since last decade, a large number of causal or closely related genes have been reported for many diseases by experimental or computational methods [84, 85]. However, a complex disease usually reflects the perturbation to the complex intracellular network, rather than a consequence of an abnormality within a single gene [86]. By studying disease genes in the context of biological networks, we consider the disease genes as a whole instead of studying them individually. Such studies may not only provide clues to uncover the molecular mechanisms of diseases, but also reveal distinguishing properties of disease genes, which can be used to predict unknown disease genes.

5.5.1 Network Characteristics Analysis of Disease Genes

Interactions among disease genes in biological networks. Disease genes can be mapped into the network (Fig. 5.9a), and a sub-network around them can be extracted to obtain a view of the local interactions among them [27]. It is well-known that the protein products of different genes harboring causal mutations for the same Mendelian disease often physically interact. A recent study suggested that in many complex diseases, proteins encoded by genes from disease-associated regions also tend to physically interact [87]. This characteristic is the foundation of "guilty-by association" policy to predict unknown disease genes.

Distinct network properties of disease genes. Studies have found that some network properties can distinguish a group of disease genes from background genes or another set of genes, and thus are particularly informative for the relevant disease (Fig. 5.9b, c). In yeast, it was found that disease genes in general tend to have higher degrees, cluster together, and locate at the central network locations [88], but another study on human did not find higher degrees for disease genes [89]. In humans, it was reported that cancer proteins tend to have higher degrees and locate at central part of the network [90]. Moreover, it was found that cancer proteins tend to have higher betweenness (which measures the importance of a gene in communication between other gene pairs) and shorter shortest-paths than both the essential and the background proteins [91]. The specificity of network characteristics of disease genes can provide us clues to specific mechanisms behind the diseases.

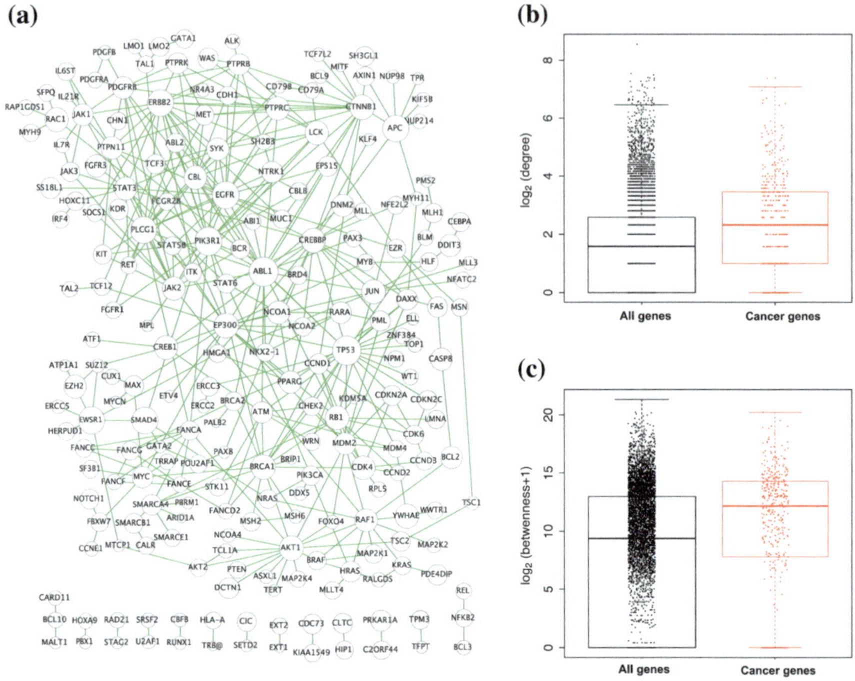

Fig. 5.9 Network characteristics of cancer genes. Among 547 cancer genes from COSMIC (Version 70; Aug 2014) [112], 386 of them were analyzed in the background network HINT [113]. **a** 394 directly physical interactions between cancer genes products. **b** Cancer genes tend to have higher degrees than background genes in HINT ($P = 5.136 \times 10^{-22}$, Wilcoxon rank-sum test). **c** Cancer genes tend to have higher betweenness than background genes in HINT ($P = 3.509 \times 10^{-18}$, Wilcoxon rank-sum test)

Network characteristics of disease genes in different biological networks and species. A recent cancer study found that prognostic genes are less likely to be hub genes in co-expression networks, and this pattern is unique to the corresponding cancer-type-specific network. Enriched in modules, prognostic genes are especially likely to be module genes conserved across different cancer co-expression networks [92]. In addition to co-expression network, researchers also integrated tissue-specific gene expression with protein interaction to derive tissue-specific PPI networks [93]. This provides an opportunity to study network characteristics of disease genes in tissue-specific PPI networks.

5.5.2 *Software Tools for Network Characteristics Analysis*

Many software tools have been developed for network characteristics analysis (Table 5.2). Some allow users to upload their own gene list for targeted analysis.

Table 5.2 Tools for network characteristics analysis

Tools	Description and access	Ref.
Targeted analysis		
TopoGSA	Generate 2D or 3D plots of network characteristics to visualize the network characteristic for each uploaded gene. Comparison with known gene sets based on 2D or 3D plots to visually identify similar pathways to the uploaded dataset The Web server can be accessed at http://www.topogsa.org	[94]
SNOW	Compute several network characteristics and estimate the statistical significance by comparing the network characteristics of the uploaded genes to those of the background genes or those in random networks The Web server can be accessed at http://snow.bioinfo.cipf.es	[95]
NetworkAnalyzer	Compute and display a comprehensive set of topological parameters. It can analyze the whole network or subset of nodes from the network It is a plug-in of Cytoscape	[96]
General analysis		
CentiScaPe	Compute 9 kinds of centralities of genes (proteins) in biological networks. It can highlight the genes whose centralities are higher or (lower) than the user-defined thresholds. It can generate "plot by node," which shows the centralities of one gene with background information about the centralities (e.g., min, mean). It can also generate "plot by centrality" to identify group of genes clustered together according to combinations of centralities. Attributes from experiments can be also uploaded to analyze relationship between experimental data and gene centralities It is a plug-in of Cytoscape	[108]
CenTiBiN	Compute and explore 17 kinds of centralities of genes (proteins) in biological networks The Web server can be accessed at http://centibin.ipk-gatersleben.de, and there is also instable Windows application.	[109]
CentiLib	CentiLib is a Java-based library and user-friendly plug-in for the analysis and visual exploration of centralities in networks. CentiLib can achieve similar functions as CenTiBiN, but it is easier to use and it can deal with weighted networks The software and manual can be downloaded at http://centilib.ipk-gatersleben.de/	[110]

For example, TopoGSA can generate 2D or 3D plots for submitted genes, which show difference network characteristics simultaneously [94]. When microarray data are uploaded, differentially expressed genes can be automatically identified and used as targeted genes for the analysis. TopoGSA can also compare the network characteristics of targeted genes with those of known gene sets (e.g., pathways). SNOW [95], a similar tool, can calculate the network characteristics and estimate their statistical significance. NetworkAnalyzer can also carry out a similar analysis when genes from the network are selected [96]. In addition to these methods, several tools for general network analysis can also be helpful (Table 5.2).

5.5.3 Association Between Disease Genes and Other Gene Sets

Another important utility of networks is to find the association between disease genes and other functional groups of genes. For example, recent studies suggested that the it is important to consider the relationship between genetic diseases and the aging process for understanding the molecular mechanisms of complex diseases. To better understand such association, one study investigated the relationship among aging genes and disease genes in a human disease-aging network [97]. The study found that (1) human disease genes are much closer to aging genes than expected by chance; (2) aging genes contribute significantly to association among diseases compared with nonaging genes with similar degrees.

It is important to assess functional association between a group of genes (e.g., candidate disease genes) and predefined gene sets. Overrepresentation-based enrichment analysis is commonly used for this task. This method, however, has several shortcomings. First, only shared genes between the input gene list and the known gene sets are considered, but current data of gene sets are not complete. Second, genes in the gene sets are treated equally, disregarding the network structure of physical or functional interactions between genes. To address these limitations, it is applicable to combine information of protein–protein interaction network with known get sets. To tackle these problems, several such tools have been developed. Glaab et al. [98] combined information from pathways databases and interaction networks and obtained more robust pathways and process representations. Their method first maps the genes in pathways into a protein–protein interaction network and then extends the pathways by including densely interacting partners. Later, Glaab et al. [24] proposed another tool for network-based gene set enrichment analysis. This approach first maps the target genes and reference gene sets into the network. It then scores the distance between the mapped target genes and reference dataset using a random walk with restart algorithm and compares the score against a background model. This method can use the network distance to differentiate gene sets with similar enrichment levels assessed by overrepresentation analysis. More importantly, it can identify novel functional associations (with no or few shared genes) and can evaluate tissue-specific association.

5.6 Conclusions

Gene expression is under tight regulation at all levels in normal cells. The characteristic forms and behaviors of different cell types are the result of their varying patterns of expression of the same set of genes. The dysregulation of gene expression can cause abnormal cell behaviors and result in diseases, and thus, gene expression profiling could provide the first clue about the molecular mechanisms of a disease. Two recent developments are spearheading the advancement of disease research in this field: First, next-generation sequencing technologies have increased the throughput and the resolution of gene expression studies to an unprecedented level; second, new computational methods with sophisticated data integration, especially network integration, have been developed for gene expression data analysis. Biological networks can provide important a priori functional information in data analysis, and since last decade, many different types of them have been constructed: Not only the number has increased but also the coverage of them has increased dramatically. With such recent resource and technology development, biology has entered a new data-driven phase in the twenty-first century. Now is a particularly challenging and exciting time for disease research with gene expression assay, as more and more gene expression data are being generated at an ever-accelerating speed.

References

1. Manel E, Paul C, Stephen B, James H. A gene hypermethylation profile of human cancer. Cancer Res. 2001;61:3225–9.
2. Jones PA, Baylin SB. The fundamental role of epigenetic events in cancer. Nat Rev Genet. 2002;3:415–28.
3. Darnell JE Jr. Transcription factors as targets for cancer therapy. Nat Rev Cancer. 2002;2:740–9.
4. Hanahan D, Weinberg RA. Hallmarks of cancer: the next generation. Cell. 2011;144:646–74.
5. Herz HM, Hu D, Shilatifard A. Enhancer malfunction in cancer. Mol Cell. 2014;53:859–66.
6. Croce CM. Causes and consequences of microRNA dysregulation in cancer. Nat Rev Genet. 2009;10:704–14.
7. Akavia UD, Litvin O, Kim J, Sanchez-Garcia F, Kotliar D, et al. An integrated approach to uncover drivers of cancer. Cell. 2010;143:1005–17.
8. Kim YA, Wuchty S, Przytycka TM. Identifying causal genes and dysregulated pathways in complex diseases. PLoS Comput Biol. 2011;7:e1001095.
9. Herranz H, Cohen SM. MicroRNAs and gene regulatory networks: managing the impact of noise in biological systems. Genes Dev. 2010;24:1339–44.
10. Tsang J, Zhu J, van Oudenaarden A. MicroRNA-mediated feedback and feedforward loops are recurrent network motifs in mammals. Mol Cell. 2007;26:753–67.
11. Stark A, Brennecke J, Bushati N, Russell RB, Cohen SM. Animal MicroRNAs confer robustness to gene expression and have a significant impact on 3'UTR evolution. Cell. 2005;123:1133–46.
12. Lu M, Zhang Q, Deng M, Miao J, Guo Y, et al. An analysis of human microRNA and disease associations. PLoS ONE. 2008;3:e3420.
13. Ramos RG, Olden K. Gene-environment interactions in the development of complex disease phenotypes. Int J Environ Res Public Health. 2008;5:4–11.
14. Hamosh A, Scott AF, Amberger JS, Bocchini CA, McKusick VA. Online Mendelian Inheritance in Man (OMIM), a knowledgebase of human genes and genetic disorders. Nucleic Acids Res. 2005;33:D514–7.
15. Risch N, Merikangas K. The future of genetic studies of complex human diseases. Science. 1996;273:1516–7.
16. Mayeux R. Mapping the new frontier: complex genetic disorders. J Clin Invest. 2005;115:1404–7.

17. Botstein D, Risch N. Discovering genotypes underlying human phenotypes: past successes for mendelian disease, future approaches for complex disease. Nat Genet. 2003;33 (Suppl):228–37.
18. Ritchie GR, Dunham I, Zeggini E, Flicek P. Functional annotation of noncoding sequence variants. Nat Methods. 2014;11:294–6.
19. Raney BJ, Cline MS, Rosenbloom KR, Dreszer TR, Learned K, et al. ENCODE whole-genome data in the UCSC genome browser (2011 update). Nucleic Acids Res. 2011;39:D871–5.
20. Li C, Li H. Network-constrained regularization and variable selection for analysis of genomic data. Bioinformatics. 2008;24:1175–82.
21. Zhang W, Wan YW, Allen GI, Pang K, Anderson ML, et al. Molecular pathway identification using biological network-regularized logistic models. BMC Genom. 2013;14 (Suppl 8):S7.
22. Wu C, Zhu J, Zhang X. Network-based differential gene expression analysis suggests cell cycle related genes regulated by E2F1 underlie the molecular difference between smoker and non-smoker lung adenocarcinoma. BMC Bioinform. 2013;14:365.
23. Ruan J, Dean AK, Zhang W. A general co-expression network-based approach to gene expression analysis: comparison and applications. BMC Syst Biol. 2010;4:8.
24. Glaab E, Baudot A, Krasnogor N, Schneider R, Valencia A. EnrichNet: network-based gene set enrichment analysis. Bioinformatics. 2012;28:i451–7.
25. Xia J, Gill EE, Hancock RE. NetworkAnalyst for statistical, visual and network-based meta-analysis of gene expression data. Nat Protoc. 2015;10:823–44.
26. Fryer RM, Randall J, Yoshida T, Hsiao LL, Blumenstock J, et al. Global analysis of gene expression: methods, interpretation, and pitfalls. Exp Nephrol. 2002;10:64–74.
27. Lemetre C, Zhang Q, Zhang ZD. SubNet: a Java application for subnetwork extraction. Bioinformatics. 2013;29:2509–11.
28. Marko NF, Weil RJ. Mathematical modeling of molecular data in translational medicine: theoretical considerations. Sci Transl Med. 2010;2:56rv54.
29. Wan YW, Nagorski J, Allen GI, Li ZH, Liu ZD. Identifying cancer biomarkers through a network regularized Cox model. In: Genomic Signal Processing and Statistics (GENSIPS), 2013 IEEE international workshop on IEEE. Houston, TX, 2013; pp. 36–39.
30. Wang Y, Klijn JG, Zhang Y, Sieuwerts AM, Look MP, et al. Gene-expression profiles to predict distant metastasis of lymph-node-negative primary breast cancer. Lancet. 2005;365:671–9.
31. van't Veer LJ, Dai HY, van de Vijver MJ, He YDD, Hart AAM, et al. Gene expression profiling predicts clinical outcome of breast cancer. Nature. 2002;415:530–6.
32. Parker JS, Mullins M, Cheang MC, Leung S, Voduc D, et al. Supervised risk predictor of breast cancer based on intrinsic subtypes. J Clin Oncol. 2009;27:1160–7.
33. Atias N, Istrail S, Sharan R. Pathway-based analysis of genomic variation data. Curr Opin Genet Dev. 2013;23:622–6.
34. Chuang HY, Lee E, Liu YT, Lee D, Ideker T. Network-based classification of breast cancer metastasis. Mol Syst Biol. 2007;3:140.
35. Warde-Farley D, Donaldson SL, Comes O, Zuberi K, Badrawi R, et al. The GeneMANIA prediction server: biological network integration for gene prioritization and predicting gene function. Nucleic Acids Res. 2010;38:W214–20.
36. Saito R, Smoot ME, Ono K, Ruscheinski J, Wang PL, et al. A travel guide to Cytoscape plugins. Nat Methods. 2012;9:1069–76.
37. Lee RC, Feinbaum RL, Ambros V. The C. elegans heterochronic gene lin-4 encodes small RNAs with antisense complementarity to lin-14. Cell. 1993;75:843–54.
38. Zhao Y, Ransom JF, Li A, Vedantham V, von Drehle M, et al. Dysregulation of cardiogenesis, cardiac conduction, and cell cycle in mice lacking miRNA-1-2. Cell. 2007;129:303–17.

39. Trang P, Wiggins JF, Daige CL, Cho C, Omotola M, et al. Systemic delivery of tumor suppressor microRNA mimics using a neutral lipid emulsion inhibits lung tumors in mice. Mol Ther. 2011;19:1116–22.
40. Wahlquist C, Jeong D, Rojas-Munoz A, Kho C, Lee A, et al. Inhibition of miR-25 improves cardiac contractility in the failing heart. Nature. 2014;508:531–5.
41. Ludwig N, Nourkami-Tutdibi N, Backes C, Lenhof HP, Graf N, et al. Circulating serum miRNAs as potential biomarkers for nephroblastoma. Pediatr Blood Cancer. 2015;62:1360–1367.
42. van Schooneveld E, Wildiers H, Vergote I, Vermeulen PB, Dirix LY, et al. Dysregulation of microRNAs in breast cancer and their potential role as prognostic and predictive biomarkers in patient management. Breast Cancer Res. 2015;17:526.
43. Knezevic J, Pfefferle AD, Petrovic I, Greene SB, Perou CM, et al. Expression of miR-200c in claudin-low breast cancer alters stem cell functionality, enhances chemosensitivity and reduces metastatic potential. Oncogene. 2015; doi:10.1038/onc.2015.48.
44. Langfelder P, Horvath S. WGCNA: an R package for weighted correlation network analysis. Bmc Bioinform. 2008;9:559.
45. McKinney-Freeman S, Cahan P, Li H, Lacadie SA, Huang HT, et al. The transcriptional landscape of hematopoietic stem cell ontogeny. Cell Stem Cell. 2012;11:701–14.
46. Miller JA, Ding SL, Sunkin SM, Smith KA, Ng L, et al. Transcriptional landscape of the prenatal human brain. Nature. 2014;508:199.
47. Okamura Y, Aoki Y, Obayashi T, Tadaka S, Ito S, et al. COXPRESdb in 2015: coexpression database for animal species by DNA-microarray and RNAseq-based expression data with multiple quality assessment systems. Nucleic Acids Res. 2015;43:D82–6.
48. Marbach D, Costello JC, Kuffner R, Vega NM, Prill RJ, et al. Wisdom of crowds for robust gene network inference. Nat Methods. 2012;9:796.
49. Faith JJ, Hayete B, Thaden JT, Mogno I, Wierzbowski J, et al. Large-scale mapping and validation of Escherichia coli transcriptional regulation from a compendium of expression profiles. PLoS Biol. 2007;5:54–66.
50. Huynh-Thu VA, Irrthum A, Wehenkel L, Geurts P. Inferring regulatory networks from expression data using tree-based methods. PLoS One. 2010;5:e12776.
51. Salgado H, Peralta-Gil M, Gama-Castro S, Santos-Zavaleta A, Muniz-Rascado L, et al. RegulonDB v8.0: omics data sets, evolutionary conservation, regulatory phrases, cross-validated gold standards and more. Nucleic Acids Res. 2013;41:D203–13.
52. Jiang J, Gusev Y, Aderca I, Mettler TA, Nagorney DM, et al. Association of MicroRNA expression in hepatocellular carcinomas with hepatitis infection, cirrhosis, and patient survival. Clin Cancer Res. 2008;14:419–27.
53. Jopling CL, Yi M, Lancaster AM, Lemon SM, Sarnow P. Modulation of hepatitis C virus RNA abundance by a liver-specific MicroRNA. Science. 2005;309:1577–81.
54. Wang X, Zhang X, Ren XP, Chen J, Liu H, et al. MicroRNA-494 targeting both proapoptotic and antiapoptotic proteins protects against ischemia/reperfusion-induced cardiac injury. Circulation. 2010;122:1308–18.
55. Xu J, Hu Z, Xu Z, Gu H, Yi L, et al. Functional variant in microRNA-196a2 contributes to the susceptibility of congenital heart disease in a Chinese population. Hum Mutat. 2009;30:1231–6.
56. Abelson JF, Kwan KY, O'Roak BJ, Baek DY, Stillman AA, et al. Sequence variants in SLITRK1 are associated with Tourette's syndrome. Science. 2005;310:317–20.
57. Yang F, Wang W, Zhou C, Xi W, Yuan L, et al. MiR-221/222 promote human glioma cell invasion and angiogenesis by targeting TIMP2. Tumour Biol. 2015;36:3763.
58. Zhao S, Yao D, Chen J, Ding N, Ren F. MiR-20a promotes cervical cancer proliferation and metastasis in vitro and in vivo. PLoS ONE. 2015;10:e0120905.
59. Houbaviy HB, Murray MF, Sharp PA. Embryonic stem cell-specific MicroRNAs. Dev Cell. 2003;5:351–8.
60. Enright AJ, John B, Gaul U, Tuschl T, Sander C, et al. MicroRNA targets in Drosophila. Genome Biol. 2003;5:R1.

61. Kertesz M, Iovino N, Unnerstall U, Gaul U, Segal E. The role of site accessibility in microRNA target recognition. Nat Genet. 2007;39:1278–84.
62. Thadani R, Tammi MT. MicroTar: predicting microRNA targets from RNA duplexes. BMC Bioinform. 2006;7(Suppl 5):S20.
63. Stingo FC, Chen YA, Vannucci M, Barrier M, Mirkes PE. A Bayesian graphical modeling approach to microRNA regulatory network inference. Ann Appl Stat. 2010;4:2024–48.
64. Tabas-Madrid D, Muniategui A, Sanchez-Caballero I, Martinez-Herrera DJ, Sorzano CO, et al. Improving miRNA-mRNA interaction predictions. BMC Genom. 2014;15(Suppl 10):S2.
65. Ritchie W, Flamant S, Rasko JE. Predicting microRNA targets and functions: traps for the unwary. Nat Methods. 2009;6:397–8.
66. Dweep H, Sticht C, Pandey P, Gretz N. miRWalk–database: prediction of possible miRNA binding sites by "walking" the genes of three genomes. J Biomed Inform. 2011;44:839–47.
67. Xiao F, Zuo Z, Cai G, Kang S, Gao X, et al. miRecords: an integrated resource for microRNA-target interactions. Nucleic Acids Res. 2009;37:D105–10.
68. Huang JC, Babak T, Corson TW, Chua G, Khan S, et al. Using expression profiling data to identify human microRNA targets. Nat Methods. 2007;4:1045–9.
69. Joung JG, Hwang KB, Nam JW, Kim SJ, Zhang BT. Discovery of microRNA-mRNA modules via population-based probabilistic learning. Bioinformatics. 2007;23:1141–7.
70. Muniategui A, Nogales-Cadenas R, Vazquez M, Aranguren XL, Agirre X, et al. Quantification of miRNA-mRNA interactions. PLoS ONE. 2012;7:e30766.
71. Tran DH, Satou K, Ho TB. Finding microRNA regulatory modules in human genome using rule induction. BMC Bioinform. 2008;9(Suppl 12):S5.
72. Bleazard T, Lamb JA, Griffiths-Jones S Bias in microRNA functional enrichment analysis. Bioinformatics. 2015;31:1592–1598.
73. Gusev Y, Schmittgen TD, Lerner M, Postier R, Brackett D. Computational analysis of biological functions and pathways collectively targeted by co-expressed microRNAs in cancer. BMC Bioinform. 2007;8(Suppl 7):S16.
74. Liu B, Li J, Cairns MJ. Identifying miRNAs, targets and functions. Brief Bioinform. 2014;15:1–19.
75. Welter D, MacArthur J, Morales J, Burdett T, Hall P, et al. The NHGRI GWAS Catalog, a curated resource of SNP-trait associations. Nucleic Acids Res. 2014;42:D1001–6.
76. Schizophrenia Working Group of the Psychiatric Genomics C. Biological insights from 108 schizophrenia-associated genetic loci. Nature. 2014;511:421–7.
77. Nicolae DL, Gamazon E, Zhang W, Duan S, Dolan ME, et al. Trait-associated SNPs are more likely to be eQTLs: annotation to enhance discovery from GWAS. PLoS Genet. 2010;6:e1000888.
78. Andersson R, Gebhard C, Miguel-Escalada I, Hoof I, Bornholdt J, et al. An atlas of active enhancers across human cell types and tissues. Nature. 2014;507:455–61.
79. Maurano MT, Humbert R, Rynes E, Thurman RE, Haugen E, et al. Systematic localization of common disease-associated variation in regulatory DNA. Science. 2012;337:1190–5.
80. Hou L, Zhao H. A review of post-GWAS prioritization approaches. Front Genet. 2013;4:280.
81. Thurman RE, Rynes E, Humbert R, Vierstra J, Maurano MT, et al. The accessible chromatin landscape of the human genome. Nature. 2012;489:75–82.
82. Sullivan PF, Lin D, Tzeng JY, van den Oord E, Perkins D, et al. Genomewide association for schizophrenia in the CATIE study: results of stage 1. Mol Psychiatry. 2008;13:570–84.
83. Kelemen O, Kovacs T, Keri S. Contrast, motion, perceptual integration, and neurocognition in schizophrenia: the role of fragile-X related mechanisms. Prog Neuropsychopharmacol Biol Psychiatry. 2013;46:92–7.
84. Bauer-Mehren A, Bundschus M, Rautschka M, Mayer MA, Sanz F, et al. Gene-disease network analysis reveals functional modules in mendelian, complex and environmental diseases. Plos One. 2011;6:e20284.
85. Rappaport N, Nativ N, Stelzer G, Twik M, Guan-Golan Y, et al. MalaCards: an integrated compendium for diseases and their annotation. Database J Biol Databases Curation. 2013; bat018.

86. Barabasi AL, Gulbahce N, Loscalzo J. Network medicine: a network-based approach to human disease. Nat Rev Genet. 2011;12:56–68.
87. Rossin EJ, Lage K, Raychaudhuri S, Xavier RJ, Tatar D, et al. Proteins encoded in genomic regions associated with immune-mediated disease physically interact and suggest underlying biology. Plos Genet. 2011;7:e1001273.
88. Ideker T, Sharan R. Protein networks in disease. Genome Res. 2008;18:644–52.
89. Goh KI, Cusick ME, Valle D, Childs B, Vidal M, et al. The human disease network. Proc Natl Acad Sci USA. 2007;104:8685–90.
90. Jonsson PF, Bates PA. Global topological features of cancer proteins in the human interactome. Bioinformatics. 2006;22:2291–7.
91. Sun JC, Zhao ZM. A comparative study of cancer proteins in the human protein-protein interaction network. Bmc Genomics 2010;11.
92. Yang Y, Han L, Yuan Y, Li J, Hei NN, et al. Gene co-expression network analysis reveals common system-level properties of prognostic genes across cancer types. Nat Commun 2014;5:3231.
93. Magger O, Waldman YY, Ruppin E, Sharan R. Enhancing the prioritization of disease-causing genes through tissue specific protein interaction networks. Plos Comput Biol 2012;8:e1002690.
94. Glaab E, Baudot A, Krasnogor N, Valencia A. TopoGSA: network topological gene set analysis. Bioinformatics. 2010;26:1271–2.
95. Minguez P, Gotz S, Montaner D, Al-Shahrour F, Dopazo J. SNOW, a web-based tool for the statistical analysis of protein-protein interaction networks. Nucleic Acids Res. 2009;37: W109–14.
96. Assenov Y, Ramirez F, Schelhorn SE, Lengauer T, Albrecht M. Computing topological parameters of biological networks. Bioinformatics. 2008;24:282–4.
97. Wang JG, Zhang SH, Wang Y, Chen LN, Zhang XS. Disease-aging network reveals significant roles of aging genes in connecting genetic diseases. Plos Comput Biology 2009;5: e1000521.
98. Glaab E, Baudot A, Krasnogor N, Valencia A. Extending pathways and processes using molecular interaction networks to analyse cancer genome data. Bmc Bioinform. 2010;11:597.
99. Kozomara A, Griffiths-Jones S. miRBase: annotating high confidence microRNAs using deep sequencing data. Nucleic Acids Res. 2014;42:D68–73.
100. Betel D, Wilson M, Gabow A, Marks DS, Sander C. The microRNA.org resource: targets and expression. Nucleic Acids Res. 2008;36:D149–53.
101. Ru Y, Kechris KJ, Tabakoff B, Hoffman P, Radcliffe RA, et al. The multiMiR R package and database: integration of microRNA-target interactions along with their disease and drug associations. Nucleic Acids Res. 2014;42.e133.
102. Jacobsen A, Silber J, Harinath G, Huse JT, Schultz N, et al. Analysis of microRNA-target interactions across diverse cancer types. Nat Struct Mol Biol. 2013;20:1325–32.
103. Jiang Q, Wang Y, Hao Y, Juan L, Teng M, et al. miR2Disease: a manually curated database for microRNA deregulation in human disease. Nucleic Acids Res. 2009;37:D98–104.
104. Purcell SM, Moran JL, Fromer M, Ruderfer D, Solovieff N, et al. A polygenic burden of rare disruptive mutations in schizophrenia. Nature. 2014;506:185–90.
105. Xie B, Ding Q, Han H, Wu D. miRCancer: a microRNA-cancer association database constructed by text mining on literature. Bioinformatics. 2013;29:638–44.
106. Bhattacharya A, Ziebarth JD, Cui Y. PolymiRTS Database 3.0: linking polymorphisms in microRNAs and their target sites with human diseases and biological pathways. Nucleic Acids Res. 2014;42:D86–91.
107. Bruno AE, Li L, Kalabus JL, Pan Y, Yu A, et al. miRdSNP: a database of disease-associated SNPs and microRNA target sites on 3'UTRs of human genes. BMC Genom. 2012;13:44.
108. Scardoni G, Petterlini M, Laudanna C. Analyzing biological network parameters with CentiScaPe. Bioinformatics. 2009;25:2857–9.

109. Junker BH, Koschutzki D, Schreiber F. Exploration of biological network centralities with CentiBiN. Bmc Bioinform. 2006;7:219.
110. Grassler J, Koschutzki D, Schreiber F. CentiLib: comprehensive analysis and exploration of network centralities. Bioinformatics. 2012;28:1178–9.
111. Hindorff LA MJEBI, Morales J (European Bioinformatics Institute), Junkins HA, Hall PN, Klemm AK, Manolio TA. (Available at: http://www.genome.gov/gwastudies). A catalog of published genome-wide association studies. Accessed 31 Mar 2015.
112. Forbes SA, Beare D, Gunasekaran P, Leung K, Bindal N, et al. COSMIC: exploring the world's knowledge of somatic mutations in human cancer. Nucleic Acids Res. 2014;43: D805–11.
113. Das J, Yu HY. HINT: High-quality protein interactomes and their applications in understanding human disease. BMC Syst Biol. 2012;6:92.

Author Biographies

Dr. Quanwei Zhang received his PhD degree at Xi'an Jiaotong University in China in 2010, after training in machine learning and bioinformatics. He joined the bioinformatics core in Northwestern University as a postdoc from 2011 to 2013, where his research focused on statistical and computational analysis of next-generation sequencing data, including nucleosome-positioning, histone methylation, and protein-binding sites analysis. He joined the Division of Computational Genetics at Albert Einstein College of Medicine in 2013 as a research fellow. Since then, he has been working on human aging. By integrating biological networks and the human genetic variation together, he is looking for clues to the basic mechanisms of aging and the evolution of sub-network of aging genes.

Dr. Wen Zhang obtained his PhD in computational science from Shanghai University in 2009. Afterward, he studied at Michigan Technological University where he got his second MS degree, in bioinformatics, in 2012. After working at Baylor College of Medicine for one year, he joined Prof. Zhengdong Zhang's laboratory at Albert Einstein College of Medicine in 2013, as a postdoctoral fellow working on bioinformatics and computational genetics. He is currently analyzing high-throughput data from human genome sequencing and genome-wide association studies.

Dr. Rubén Nogales-Cadenas is a scientific researcher at the Albert Einstein College of Medicine in New York. He has more than eight years of research experience in Bioinformatics and Computational Biology. His scientific activity is mainly focused on the integration of biological data and computational resources for genomics research of the immune system, cancer, and drug discovery. He is interested in a wide range of research topics, including the development and implementation of new data analysis methodologies and their application to real biological problems, always taking advantage of many different types of biological information from experimental, functional, structural, regulatory, and pharmacological data. Dr. Nogales-Cadenas has published 18 articles in international well-recognized journals.

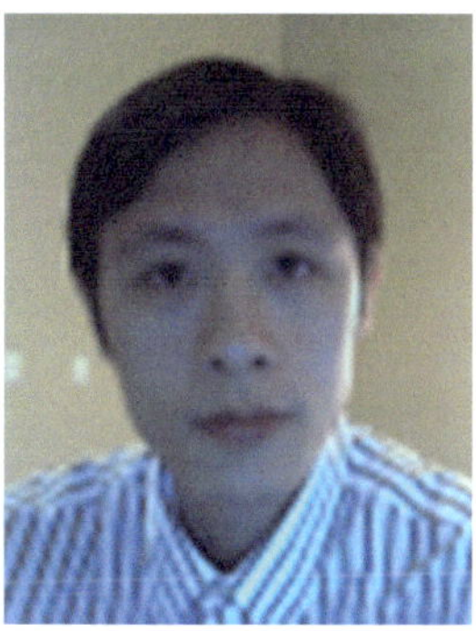

Dr. Jhih-Rong Lin received his PhD degree in computer science at University of South Carolina in 2013 with the thesis on computational analysis of protein sorting signals and localization. In 2014, he joined the Genetics Department of Albert Einstein College of Medicine as a research fellow. Currently, he is working on prediction of causal genes and causal variants of complex diseases. His main research interest focuses on the development of computational methods for human genetic analysis. As the first author of three journal articles in the field of bioinformatics, he has developed free software tools implementing his computational methods for public access.

Ying Cai obtained her MS degree in Medical Genome Sciences from the University of Tokyo, where she studied miRNA expression in adult T-cell leukemia. As a PhD candidate in the Department of Genetics at Albert Einstein College of Medicine, she is currently working in the field of bioinformatics, focusing on the analysis of mRNA expression profile of breast cancer.

Prof. Zhengdong D. Zhang is an assistant professor in the Department of Genetics at Albert Einstein College of Medicine. His research interests are computational genomics and systems biology of complex human diseases, focusing on algorithm development, data integration, and software implementation (visit www.zdzlab.org for more information). He participated and played an active role in some of the most notable international genome projects. He has investigated human functional genomics on different levels—from single genes, to gene families, and to the whole genomes—with an integrative approach drawing form molecular biology, statistics, and computational biology. At the Baylor College of Medicine Human Genome Sequencing Center, as part of the Rat Genome Project, he performed a comparative analysis of the nuclear receptor family in the human, mouse, and rat. At Yale University, as part of the ENCODE and the 1000 Genomes Projects, he developed software pipelines to process microarray and high-throughput sequencing data and carried out a detailed statistical analysis of the genomic distribution and correlation of transcriptional regulatory elements in the ENCODE regions. At Albert Einstein College of Medicine, his research team recently developed two computational frameworks of data integration to identify risk genes of complex diseases. Using a novel framework of integrated post-GWAS analysis, they identified distal causal genes of complex human diseases linked to GWAS signals through nearby regulatory elements such as enhancers. In another study, they developed a network-regularized logistic regression method. Using it to analyze case–control sequencing data, they identified and prioritized risk genes of complex human diseases. He received the NIH Career Development Award from the National Library of Medicine and the New Scholar Award from the Ellison Medical Foundation for his scientific and medical research.

Chapter 6
Introduction to Isoform Sequencing Using Pacific Biosciences Technology (Iso-Seq)

Manuel L. Gonzalez-Garay

Abstract Alternative RNA splicing is a known phenomenon, but we still do not have a complete catalog of isoforms that explain variability in the human transcriptome. We have made significant progress in developing methods to study variability of the transcriptome, but we are far away of having a complete picture of the transcriptome. The initial methods to study gene expression were based on cloning of cDNAs and Sanger sequencing. The strategy was labor-intensive and expensive. With the development of microarrays, different methods based on exon arrays and tiling arrays provided valuable information about RNA expression. However, the microarray presented significant limitations. Most of the limitations became apparent by 2005, but it was not until 2008 that an alternative method to study the transcriptome was developed. RNA Sequencing using next-generation sequencing (RNA-Seq) quickly became the technology of choice for gene expression profiling. Recently, the precision and sensitivity of RNA-Seq have come into question, especially for transcriptome reconstruction. This chapter will describe a relatively new method, "Isoform Sequencing" (Iso-Seq). Iso-Seq was developed by Pacific Biosciences (PacBio), and it is capable of identifying new isoforms with extraordinary precision due to its long-read technology. The technique to create libraries is straightforward, and the PacBio RS II instrument generates the information in hours. The bioinformatics analysis is performed using the freely available SMRT® Portal software. The SMRT® Portal is easy to use and capable of performing all the steps necessary to analyze the raw data and to generate high-quality full-length isoforms. For the universal acceptance of the Iso-Seq method, the capacity of the SMRT® Cells needs to improve at least 10- to 100-fold to make the system affordable and attractive to users.

Keywords Isoform • Pacific biosciences • Iso-Seq • Pacbio • SMRT • RNA-Seq

M.L. Gonzalez-Garay (✉)
Center for Molecular Imaging. The Brown Foundation Institute of Molecular Medicine for the Prevention of Human Diseases, The University of Texas, Health Science Center at Houston, 1825 Pressler Street SRB 330H, Houston, TX 77030, USA
e-mail: manuel.l.gonzalezgaray@uth.tmc.edu

J. Wu, *Transcriptomics and Gene Regulation*,
Translational Bioinformatics 9, DOI 10.1007/978-94-017-7450-5_6

6.1 Introduction

The complete set of all RNA molecules in a cell or a population of cells is called the transcriptome. Qualitative and quantitative information about the transcriptome is essential to understand the molecular mechanisms of cellular physiology. The first attempts to study gene expression dated back to the late 1970s. During that period, Dr. James Alwine and colleges from Stanford University developed a new method, the Northern Blot. Northern Blot consists of running agarose gels, transferring the RNA to membranes, and hybridizing with radioactive probes, in order to detect a few genes per experiment [3]. Moving from single-gene studies to full transcriptomes required the development of catalogs of genes and RNAs. The first catalog of RNA molecules was developed using Expressed sequence tags (EST) and complementary DNA (cDNA) sequences [36, 46]. Such catalogs were used during the early 1990s to create custom-made microarrays and eventually commercial high-density oligo microarrays. A group of microarrays, the exon arrays, contained probes for all the exon from known genes, including exon/intron boundaries. These arrays were used in expression profiling and played a significant role in detecting new alternatively spliced isoforms and monitoring their expression patterns in different tissues. Another type of microarrays, the tiling arrays, contains overlapping probes for regions of the genome, especially in coding regions. The high density of probes in coding regions allowed the mapping of new genes and isoforms [33]. By 2006, issues with reliability and reproducibility of data obtained from expression microarrays were apparent in the scientific community [1, 13]. Some of the issues included: (a) limited range of sensitivity on both the low and high ends; (b) cross-hybridization and non-specific hybridization issues; (c) probe saturation, affecting the accurate quantification of high-abundance transcripts; and (d) the presence of single nucleotide polymorphisms and insertions or deletions in samples, impacting the performance of a probe [2, 6, 33, 49, 57].

6.2 The RNA-Seq Era

With the development of the next-generation sequencing technology (NGS), finding the application of this technology to RNA sequencing was a logical step. During 2008, new RNA sequencing methods were contributed by four different groups that published their protocols and their datasets within days of each other. All the four groups used the Illumina sequencing technology applied to sequence RNA from different organisms [30, 34, 35, 62]. The new method was named RNA-Seq, after the catchy name given by Mortazavi's group [34]. An interesting fact during that period was that a different group published a similar method a year before the four groups, but used pyrosequencing technology [61].

As soon as the new data were released, bioinformaticians started to develop new algorithms and software capable of analyzing the new type of information generated by RNA-Seq. By 2009, publications started to emerge reporting the development of new software such as the splice-aware aligner, Bowtie [25]; a new statistical method to estimate differential expression, DEGseq [59]; a tool to discover splice junctions, TopHat [55]; and a transcriptome reconstruction program, Cufflinks [56].

During the last six years, the original software packages matured very quickly, and a myriad of new software packages has been continuously generated by the scientific community [65].

6.2.1 The Standard RNA-Seq Protocol

Figure 6.1 shows an overview of the steps in a standard RNA-Seq workflow. The workflow consists of five basic steps:

(a) Isolation of RNA.
(b) Isolation of polyA mRNA (optional).
(c) RNA fragmentation.
(d) Synthesis of high-quality double-stranded cDNA.
(e) Library preparation.

The library is then submitted to sequencing [60]. The resulting sequencing reads are stored in one or two files in FASTQ format. FASTQ is the standard file format for DNA sequencing data, which stores both nucleotides and quality scores [11].

6.2.2 RNA-Seq Analysis

The analysis strategy falls into two categories depending on whether a reference genome assembly is available. If the reference genome is available, the best approach is to use a reference-based strategy. If the reference genome is not available, a de novo strategy should be used.

Figure 6.2 shows a typical analytical workflow for the reference-based strategy. The process consists of the following steps:

1. The raw reads (contained in FASTQ files) are mapped to a reference genome using a splice-aware aligner. Most aligners take advantage of the prior knowledge provided by the reference genome and the current information about the structure of the genes that have already been mapped to the genome (GFF annotations). Currently, most if not all the splice-aware aligner store the alignments in a file of type BAM.
2. The transcript reconstruction is performed by using the overlapping reads to create a graph that represents all possible splicing isoforms. Then full-length isoforms are generated from the graph.

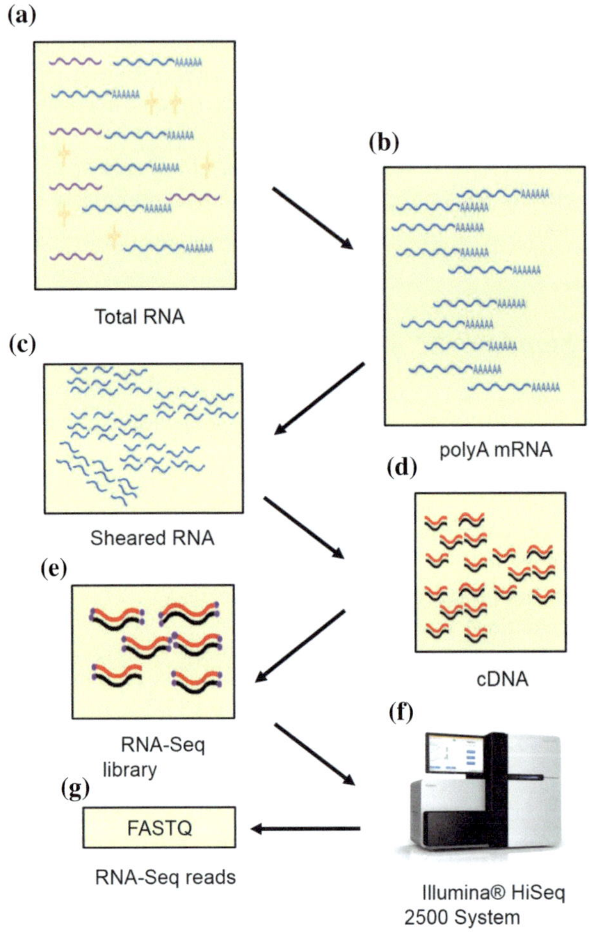

Fig. 6.1 Overview of a typical RNA-Seq protocol. Total RNA is Isolated (**a**); polyA mRNA fraction is isolated (**b**); RNA is fragmented (**c**); high-quality double-stranded cDNA is synthesized (**d**); adaptors are ligated to the cDNA to create a library (**e**); the library needs to pass a quality control, and then, it is loaded into the sequencing instrument (**f**); after few days, the instrument will generate raw information. The raw information need to be preprocessed with Illumina's proprietary software CASAVA to generate FASTQ files (**g**)

3. The majority of RNA-Seq experiments use replicas and comparisons between samples, and consequently, it is a part of the workflow to merge the results from multiple transcript assemblies to generate a global transcriptome.
4. Differential expression between samples is calculated using statistical analysis.

There are other additional steps implemented by some packages such as normalization and transcript abundance; however, for simplification, these steps are not shown in Fig. 6.2.

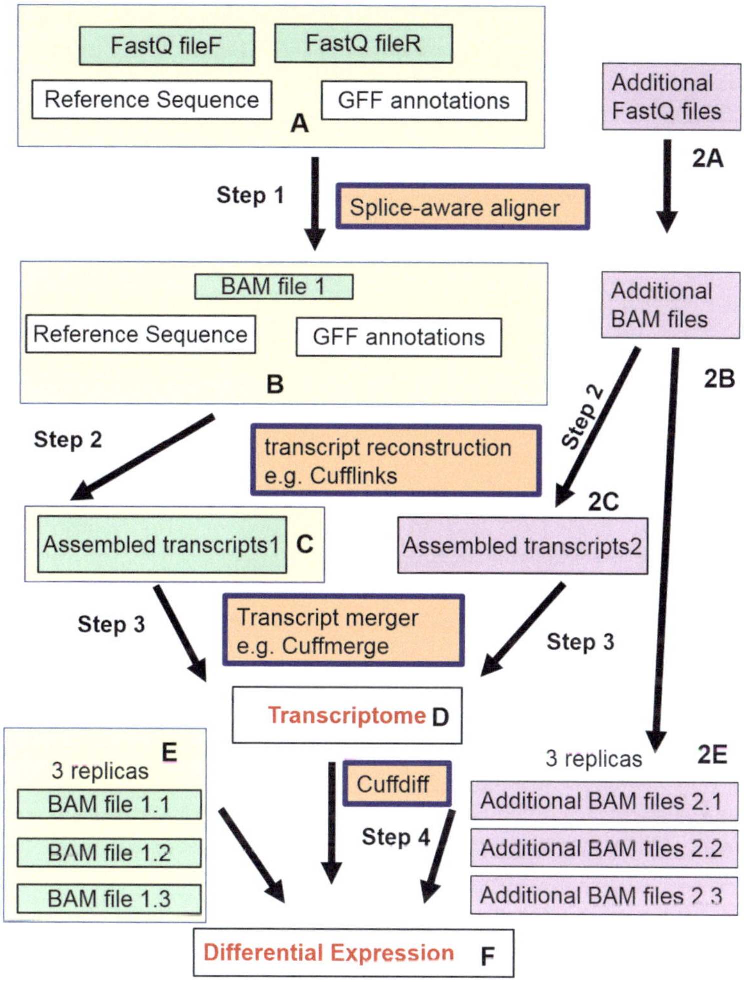

Fig. 6.2 Overview of a typical RNA-Seq analysis workflow. FASTQ files are mapped to a reference genome using a splice-aware aligner (*Step 1*). Some aligners like TopHat takes a gene annotation file in GFF/GTF format to improve the alignment. The input files for the splice-aware aligner are in *Box A*. The transcript reconstruction is performed in *step 2*. The input files required are in *Box B*. An assembled transcriptome is the output of the transcript reconstruction program, e.g., Cufflinks (*Box C*). Normally, this type of experiments requires replicas (*box E*), and for differential expression, an additional group of files from a second sample are used (*Boxes 2A, 2B, 2C* and *2E*). For a better reconstruction of the transcriptome, programs like cuffmerge aggregate all the assembled transcripts into a single transcriptome (*Box D*). Finally, a differential expression program like cuffdiff compares the alignments from sample-1 against the alignments of sample-2 using the transcriptome as the source of annotations (input *Boxes E, 2E* and *D*)

The strategy of de novo transcriptome assembly is used when a reference genome is not available or when the investigator wants to avoid mapping bias. The method consists in using de Bruijn graph to create contigs from the original reads.

Once the transcriptome is assembled, the process continues by annotating the transcripts, calculating abundance of isoforms and differential expression.

Both methods, reference-based and de novo, require considerable amount of computational resources to reconstruct the full-length transcripts from the reads [32, 65].

6.2.3 RNA Studies Under Fire

During 2013, Dr. Tal Nawy, associate editor of the scientific journal Nature, interviewed Dr. Snyder, a pioneer of deep RNA sequencing, and Dr. Tilgner, a member of Dr. Snyder's laboratory. Dr. Snyder commented about the ironic fact that during a RNA-Seq experiment, the RNA needed to be shredded before creating the sequencing library, and then, the complex transcriptome had to be assembled bioinformatically after the library was sequenced. Dr. Snyder argued that there was insufficient length in the NGS reads to generate a good transcript reconstruction. He recommended using longer reads from Pacific Biosciences (PacBio) to cover an entire transcript by a single read and avoid the arduous process of transcript reconstruction [37]. Dr. Snyder's comments coincided with the publications of two large-scale studies aimed to evaluate the software used during RNA-Seq. The studies were performed by the Genome Annotation Assessment Project (RGASP) consortium [16, 53]. The paper published by Engström et al. systematically evaluated the majority of the spliced alignment programs commonly used for RNA-Seq. In their publication, 26 mapping protocols based on 11 programs were evaluated, and major performance differences were discovered between methods on numerous benchmarks, including alignment yield, basewise accuracy, mismatch and gap placement, exon junction discovery, and suitability of alignments for transcript reconstruction. The alignment pipeline GSTRUCT (based on GSNAP), and the aligners GSNAP [63], MapSplice [58], and STAR [12] outperformed the rest of the aligners. However, these four programs also presented shortcomings; more noticeably, their output contained large number of false exon junctions. Not surprisingly, GSNAP [63] and GSTRUCT require considerable computing time when compared against STAR [16]. STAR [12] is known for its fast performance. The benchmarks from its original publication indicated that STAR [12] was orders of magnitude faster than GSNAP [63], MapSplice [58], and the popular aligner TopHat2 [12, 22, 55]. The Engström et al. study was an eye-opener for developers and users of the alignment programs. Since the publication of the results, there have been new releases for some of the programs described in the report. The new releases fixed some of the problems described in the paper. Other groups wrote new aligners to replace their previous aligners, as was the case for Dr. Salzberg's TopHat's aligner engine Bowtie [24, 25]. Dr. Salzberg's group developed a new aligner HISAT. According to the benchmarks provided in the publication, HISAT is the fastest system currently available, approximately 50 times faster than TopHat2 and 12 times faster than GSNAP, with equal or better accuracy than any other method [21].

The second RGASP's paper authored by Steijger et al. [53] evaluates the methods for transcript reconstruction. The group evaluated 25 protocols based on 14 computational methods. They found that all the protocols are capable of detecting annotated features with great confidence features in a simple organism like *C. elegans*. The confidence was reduced with a more complex organism like *D. melanogaster*. With a complex organism like humans, there was a great level of variability in the ability to detect annotated features from the transcriptome. Some of the methods, like Cufflinks [48], were able to detect 80 % of the annotated features, but only 60 % of the annotated features were correct. Other methods like SLIDE [27] obtained a 75 % detection rate and a 90 % precision rate. Velvet [64] and Trembly [Gerstein M unpublished] also had very good precision and detection rate, but many others have low scores in both precision and sensitivity. The main surprise was found when the programs tried to assemble a complete isoform. Valid isoforms were assembled for roughly half of expressed genes on average (*H. sapiens* mean 41 %, maximum 61 %; *D. melanogaster* mean 55 %, maximum 73 %; *C. elegans* mean 50 %, maximum 73 %), and for those genes with assembled isoforms, only one isoform was typically identified. Expression-level estimates also varied widely across methods, even when based on similar transcript models. The authors concluded that isoform identification is a limiting factor for RNA-Seq experiments [53].

6.2.4 *Long Reads, a Solution to the RNA-Seq Problem*

A solution to the current problem of RNA-Seq will require technology capable of providing long reads in the range of 1.5–10 kbp. The median length of human gene transcripts is about 2.5 kbp; long reads should be able to provide full-length mRNA isoforms, detect new isoforms, and bypass the transcript reconstruction process by identifying isoforms directly.

Recently, Roche announced the shutdown of the 454 pyrosequencing technology, and Life Technologies replaced ABI SOLiD technology in favor of the Ion Torrent sequencing technology. With those changes, there are only two major players left in the arena of next-generation sequencing: Illumina and Ion Torrent. Both technologies provide relatively short reads, Illumina with 100–150 bp reads, and Ion technology with 200–400 bp reads.

The so-called third-generation sequencing technology promises affordable, real-time sequencing with long reads (>5 kbp). There are at least two companies that fill the void for such technology: Oxford Nanopore and Pacific Biosciences. At the time of writing, the Oxford Nanopore technology is still in an early stage of development. There are several reports about the technology [5, 31, 47, 54], but it is not commercially available yet. During 2014, the results from the first early access to the MinION Nanopore sequencer were presented and discussed online. The experiences and comments from users look very encouraging.

PacBio has been available for several years through USA core centers. PacBio provides single molecule real-time (SMRT®) sequencing with reads greater than 10 Kb. Moreover, there are several freely available datasets [41], and PacBio's analytical software is available as open source software [45]. Having datasets and open source software provides a tremendous advantage for anyone willing to try this new technology.

6.3 Pacific Biosciences Technology

The direct observation of processive DNA polymerization was a paramount accomplishment and the basis of the SMRT® technology. To develop real-time sequencing, it was necessary to reduce the observation volume and bring down the concentrations of labeled nucleotides relevant to the enzyme. To solve the problem, PacBio created zero-mode waveguides (ZMWs). PacBio's ZMW is a tiny hole in an aluminum cladding film that guides light energy into a volume that is small in all dimensions compared to the wavelength of the light. PacBio uses ZWS to illuminate only a section of a well where a DNA polymerase is immobilized. The conditions in the ZWS enable single fluorophore detection despite the relatively high concentrations of labeled dNTP. PacBio also developed special fluorophores linked to the terminal phosphate moiety (phospholinked) of the nucleotide. When the DNA polymerase incorporates the nucleotide, the phosphodiester bond formation releases the fluorophore, generating a completely natural DNA strand. Generating unmodified DNA Strands prevents any adverse effects in the fidelity and rate of the DNA polymerase. The current PacBio RS II system is capable of processing 150,000 DNA fragments in a single SMRT® cell (Fig. 6.3), and each DNA fragment has the potential to generate a read of 20 Kb. One single run processes up to 16 SMRT® cells in less than 4 h [14, 23, 43]. PacBio technology has been used for de novo assembly of genomes [9, 15], targeted sequencing [7, 8], base modification detection [17, 18], and more recently for isoform sequencing (Iso-Seq). In the next section, we will focus exclusively on the Iso-Seq method.

6.3.1 Iso-Seq Experimental Pipeline

The Iso-Seq method is a new technique; the first application was published in 2012 [Larsen]. Figure 6.4 illustrates the current experimental pipeline. The pipeline consists of the following steps:

1. *Isolation of total RNA*. The method depends on the isolation of high-quality total RNA. The recommendations from PacBio and multiple publications are to use

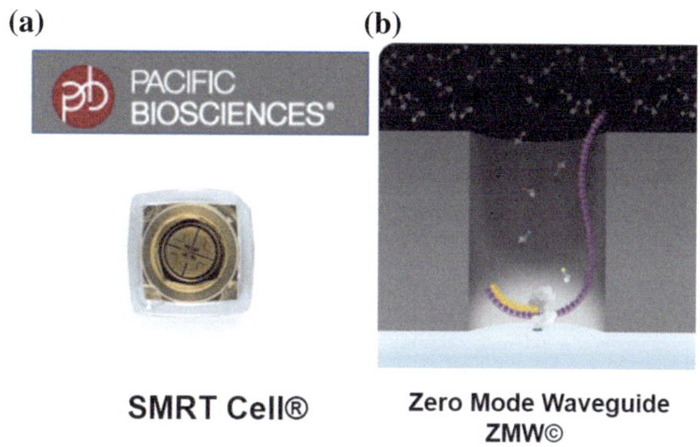

Fig. 6.3 PacBio SMRT® Cell and ZMW. The picture and the diagram were obtained from the media kit section of ©Pacific Biosciences with sole intention to explain their technology. Panel A displays a single SMRT® cell, representing the minimum unit for sequencing on a PacBio RS II. Each cell contains 150,000 ZMW, where single polymerases are immobilized and single fragments of DNA would be sequenced in real time. Panel B displays a single ZMW where one DNA polymerase is immobilized to the bottom. A laser light from the PacBio RS II illuminates the SMRT® cell from below. The light enables the detection of the release of one of four different colored fluorophores after each base is incorporated into the DNA. The final result is a real-time movie of a polymerase in action

the TRIzol® Plus RNA Purification Kit from Life Technology and to follow the manufacturer's instructions [28].

2. *PolyA mRNA selection (optional)*. The majority of the applications require the isolation of polyA mRNA, but the Iso-Seq method is flexible enough to allow the sequencing of different types of RNA.
   PacBio recommends using the Poly(A)Purist™ MAG Kit from Life Technologies for the isolation of polyA mRNA [29].
3. *cDNA synthesis with adapters*. PacBio currently recommends using the SMARTer PCR cDNA Synthesis Kit from Clontech to generate full-length cDNA [10].
4. *Size partitioning*. To avoid an over-representation of smaller transcripts, PacBio recommends using the bluePippin system [50] to fractionate the cDNA. However, it is possible to use a regular gel system or the newest sageELF system [51] to fractionate the cDNA. Size selection allows for more even representation across cDNA of different size ranges, since smaller fragments may load preferentially on the sequencer. Three different fractions are isolated: (a) 1–2 Kb; (b) 2–3 Kb; and (c) 3–6 Kb. Optionally, you could include a fourth fraction of 5–10 Kb. Furthermore, PacBio found that the larger fractions benefit from a second fractionation step to clean any carryover from the smaller fractions.
5. *Large-Scale PCR for SMRTbell™ Library Preparation*. After size selection, the double-stranded cDNA is not sufficient for SMRTbell™ library construction. PacBio recommends a PCR amplification using the KAPA HiFi Enzyme [20]. To download the entire protocol with the most updated conditions for the PCR

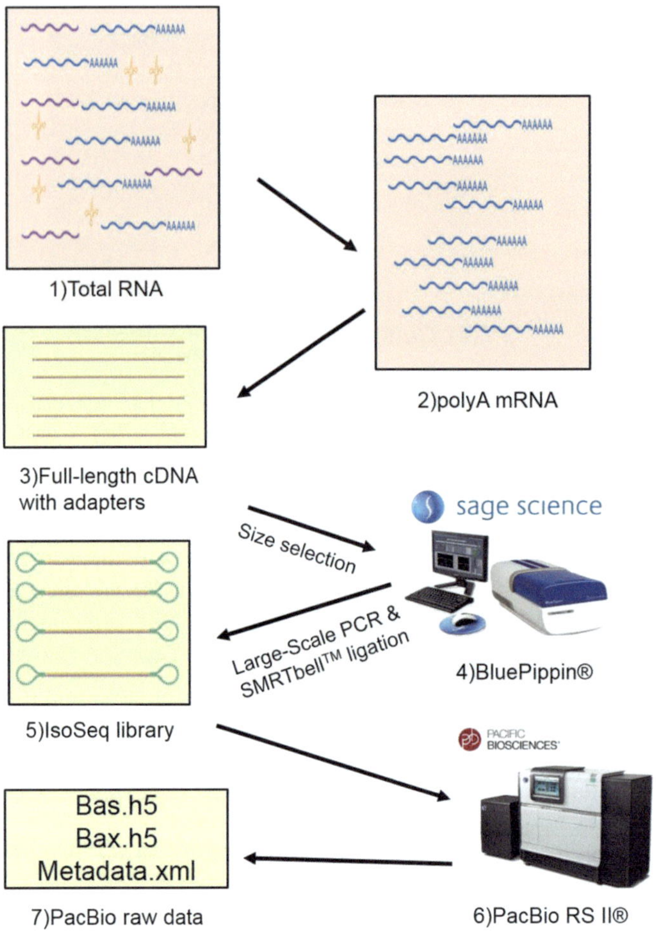

Fig. 6.4 Overview of the Iso-Seq protocol. (*1*) Total RNA is Isolated; (*2*) polyA mRNA fraction is isolated; (*3*) full-length CDNA with adapters is generated; (*4*) A size selection is performed to isolate at least three fractions: 1–2 Kb, 2–3 Kb, and 3–6 Kb. PacBio recommends to use BluePippin® system; (*5*) a large-scale PCR amplification is recommended before adding the SMRTbell™ by ligation; (*6*) the isoSeq library is loaded into the SMRT® Cell, and the cell is placed in the PacBio RS II®; and (*7*) few hours later, the raw data are available

reaction, go to the PacBio's SMRT sample prep web site [44] and search for the Iso-Seq™ procedure.

6. *SMRTbell™ Library Preparation*. The reagents necessary to transform the cDNAs into a circularized molecule are obtained directly from PacBio consumable reagents [40] and search for "SMRTbell™ Template Prep Kit 1.0." After completing this step, the library is ready to be loaded into a SMRT® Cell and placed in the instrument.

6.3.2 *Iso-Seq Analytical Pipeline*

6.3.2.1 Primary Analysis

The current instrument, PacBio RS II, includes a computational cluster "Blade Center." The blade center is responsible for primary data analysis. The primary data analysis consists of real-time signal processing, base calling, and quality assessment. Currently, each SMRT® Cell contains 150,000 ZMWs, where single polymerase is immobilized. Consequently, each SMRT® cell is capable of sequencing 150,000 DNA sequences in real time. The blade cluster performs in real time: (a) the conversion of the images (movies) to trace files; (b) the conversion of trace to pulse; (c) conversion of pulse to base; and (d) base to circular consensus. All the files generated by a single SMRT® cell are stored in a compressed archive of type HDF5 [19]. The information generated by a single SMRT® cell is stored into one bas.h5, three bax.h5 [38] files, and one metadata.xml file [39]. The metadata.xml file contains the instrument information and experimental conditions that were used during the run. It is important to notice that recently, the bas.h5 file was relocated inside the bax.h5 files; therefore, it is possible that some datasets do not have bas.h5 files present.

6.3.2.2 Secondary Analysis

The software responsible for the secondary analysis is provided by PacBio under the name "SMRT® Portal suite" (currently version 2.3). The SMRT® Portal suite is freely available from the PacBio development network [42]. The site also contains detailed information about installation. It is important to realize that there are a large number of files generated by each SMRT® cell; consequently, the software requires a computer cluster running a modern version of RedHat, CentOS, or Ubuntu. The computer cluster should have at least 16 computer nodes and a distributed computer management system like Sun Grid Engine (SGE). Moreover, SMRT® Portal will need a few terabytes of disk space to store all the raw files, temporary files, and final results. It is advisable to get help from an experienced Linux administrator to guarantee that the SMRT® suite works properly and can submit jobs to the computer cluster. The SMRT® Portal is a LAMP/Tomcat server bundled with PacBio software that functions as an interface for analyzing sequencing data generated by the PacBio RS. The SMRT® Portal can design secondary analysis jobs, submit them to the computer cluster, and generate reports from the results as soon as they are available. The SMRT® Portal user interface consists of three tabs: (a) the design job tab (Fig. 6.5a); (b) the monitor job tab; and (c) the view data tabs (Fig. 6.5b). The design job tab will import the raw data and select a protocol from a pull-down menu. Each one of the protocols contains a sequential list of calls to command line programs or scripts. PacBio developers generated some of the protocols, while users contributed others. For the Iso-Seq analysis, the RS_Iso-Seq protocol needs to be selected. Underneath the protocol name, there are submenus where parameters

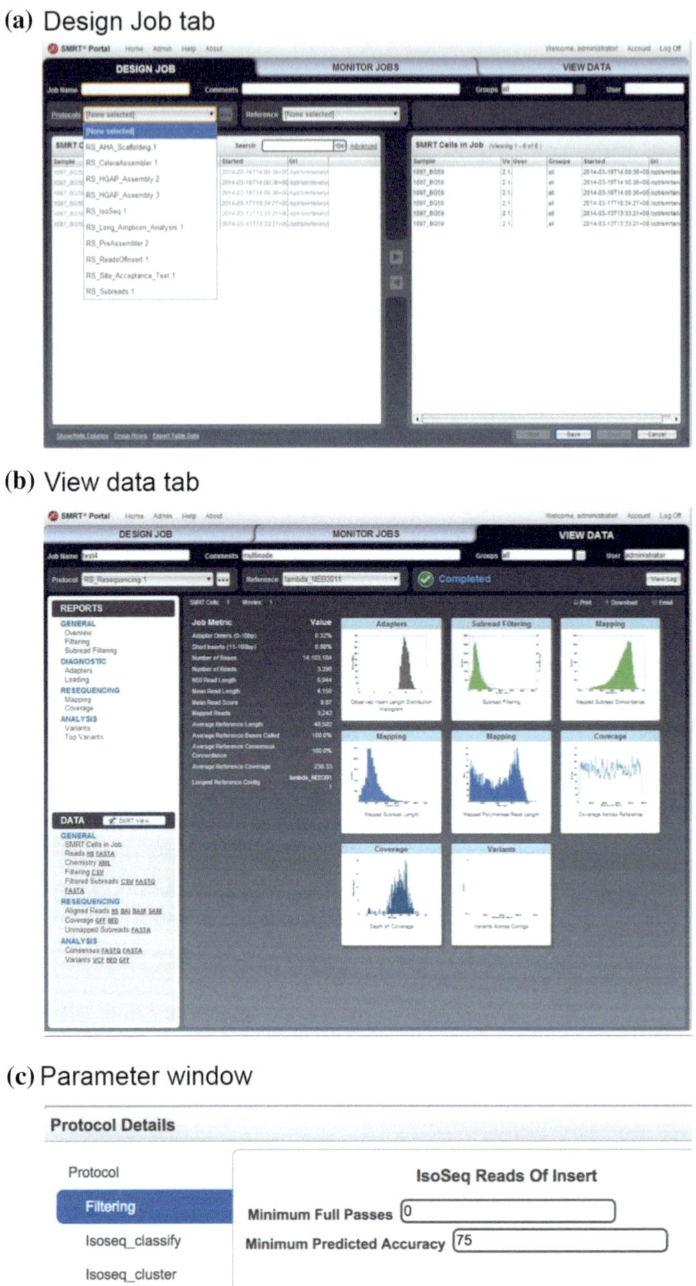
(a) Design Job tab
DESIGN JOB
MONITOR JOBS
VIEW DATA
(b) View data tab
DESIGN JOB
MONITOR JOBS
VIEW DATA
Completed
REPORTS
DATA
(c) Parameter window
Protocol Details
Protocol
Filtering
Isoseq_classify
Isoseq_cluster
IsoSeq Reads Of Insert
Minimum Full Passes
0
Minimum Predicted Accuracy
75

◀ **Fig. 6.5** The SMRT® Portal User Interface. Panel A displays the design job tab. The design job tab is the interface to import the raw data, design the job, and select the protocol. For the IsoSeq method, the RS_IsoSeq_1 protocol should be selected. Panel B shows the view data tab with an example of a completed job where you can visualize the results of the job. Panel C shows a submenu where you can set the parameters for a module

are set or modified for each one of the series of modules that belong to the protocol (Fig. 6.5c). The RS_Iso-Seq protocol requires three modules: the filtering module, the Isoseq_classifier module, and the Isoseq_cluster module. After setting all the parameters and typing a name for the job, the protocol is ready to be submitted. The job can be monitored by using the monitor job tab, and when the job is finished, the results will be displayed in the data view tab.

Figure 6.6 illustrate the Iso-Seq workflow in more detail:

(a) Reads are extracted from PacBio's raw data by the read of inserts protocol.
(b) Reads are classified as full-length and non-full-length by the classify protocol.

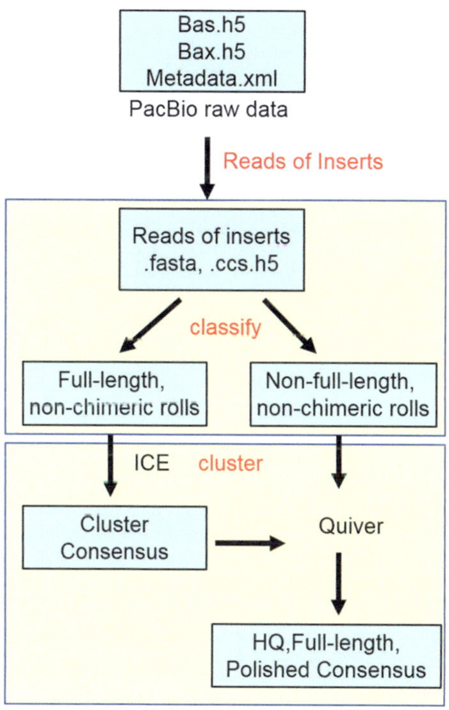

Fig. 6.6 The Iso-Seq workflow. PacBio raw data are stored in three types of files: Bas.h5, Bax.h5, and metadata.xml. After the data are imported into the SMRT® Portal, the reads of inserts module extract the reads from the raw files as.fasta files and.ccs.h5 files. The reads are classified into full-length, non-chimeric rolls and non-full-length, non-chimeric rolls by the classify module. The chimeric rolls are processed outside the SMRT® Portal interface using the developmental tools from PacBio. The full-length reads are cluster by ICE and submitted to Quiver to obtain high-quality full-length transcripts (cluster module)

(c) Full-length, non-chimeric reads are clustered using the isoform-level clustering (ICE).
(d) Isoforms from ICE are processed with Quiver to try to extend/improve the quality of the full length by using the non-full-length reads. PacBio calls the final group of full-length isoforms "Polished."

The Iso-Seq protocol does not require a reference genome to generate isoforms; however, if the reference genome is provided, the GMAP module could be activated. Alternatively, you could run GMAP form the command line. GMAP is a splice-aware aligner that maps the isoforms into the reference genome and removes redundant versions of the isoforms, cleaning up the final list of transcripts.

6.4 Scientific Applications of Iso-Seq Technology

The number of publications that use the Iso-Seq method keeps increasing every year. Since the initial development of Iso-Seq method, over a dozen articles have been published in high-impact journals. In this section, three projects will be discussed. The selected articles represent, in our opinion, important applications of the Iso-Seq method. The first paper demonstrated how the Iso-Seq method was able to identify ~50,000 different antigen-binding regions from a complex genomic region that is the target of recombination events associated with B-cell maturation [26]. The second article demonstrated the ability to identify full-length mRNAs and new isoforms in a complex transcriptome [52]. In the third paper, the authors used a hybrid approach consisting of combining Illumina and PacBio data to increase the identification of isoforms. By using statistical inference, they were able to identify over 2000 new isoforms not reported before, including 216 novel non-annotated gene loci [4].

6.4.1 Iso-Seq and Expression in Complex Genomic Regions

The Iso-Seq technology is relatively young; the first published project that used Iso-Seq was released in 2012 [26]. Larsen and Smith focused their efforts toward understanding the bovine immunoglobulin G (IgG) repertoire. The project was directed to sequence transcripts from a region with very high variability, the IgH variable region. The investigators generated ~50,000 high-quality cDNAs, from which 99.9 % corresponded to antigen-binding regions [26]. This project used a technique between amplicon sequencing and Iso-Seq sequencing, since the investigators enriched only the IgH variable region after generating the cDNA. The researchers demonstrated the power of long reads by isolating thousands of molecules that differ by a small number of variations introduced during the recombination events associated with B-cell maturation.

6.4.2 *Iso-Seq, a Tool to Use to Survey for New Isoforms*

Dr. Snyder's group demonstrated that the Iso-Seq technology is capable of sequencing full-length RNA, with little to no sequence loss at the 5′ end. In their paper, the authors were able to sequence transcripts without fragmentation or amplification. They used a commercial RNA sample that consisted of RNA from 20 different organs and tissues. The authors were able to sequence full-length RNA molecules of up to 1.5 Kb. In total, they identified ~14,000 spliced GENCODE genes that mapped with high confidence to the GENCODE annotations. Moreover, they found that 10 % of the transcripts represented new isoforms [52]. It is important to notice that the authors used the original protocol for their publication. The original protocol did not include size selection; consequently, the system preferentially sequenced smaller molecules. The current protocol discussed in the previous section used size selection to cover large isoforms. Moreover, the current protocol has additional improvements to increase yield, such as PCR optimization before size selection, and a second large-scale PCR to increase the number of rare isoforms.

6.4.3 *A Hybrid Approach Combining Illumina and PacBio Data*

In 2013, a group of scientists, led by Professor Wing Hung Wong from Stanford University, used a hybrid sequencing method to obtain a better understanding of the diversity of mRNA isoforms in human embryonic stem cells (hESCs). Their hybrid method consisted of combining Illumina sequencing information with PacBio long reads. They reported 8084 RefSeq-annotated isoforms detected as full length and an additional 5459 isoforms predicted through statistical inference. Over one-third of the predicted isoforms were novel isoforms, and they also found 275 RNAs from new gene loci. The new loci represent a group of genes expressed only in pluripotent cells. The statistical inference method used by this group is available to other researchers that wish to use their hybrid sequencing method [4].

6.5 Conclusions

In this chapter, we discussed the importance of alternative splicing and the different methods used to identify RNA isoforms. Since 2008, RNA-Seq has become the predominant technology to perform transcriptome profiling. However, recently, a team of bioinformaticians benchmarked the tools that are commonly used to analyze RNA-Seq and found significant problems when trying to assemble a transcriptome from a complex organism. None of the available tools were able to correctly recreate all the valid isoforms for a transcriptome; only half of the genes were assembled correctly, and from those genes with assembled isoforms, only one isoform was typically identified. Expression-level estimates also varied widely across methods, even when based on similar transcript models. The authors concluded that isoform identification is a limiting factor for RNA-Seq experiments [53]. Scientists like Dr. Snyder are arguing that there was insufficient length in the current NGS reads to generate good transcript reconstruction. He recommends the use of longer reads to cover the entire transcript by a single read and avoid the arduous process of transcript reconstruction. The Iso-Seq method, recently developed by PacBio, is capable of identifying new isoforms with extraordinary precision due to its long-read technology. The technique to create libraries is straightforward, and the PacBio RS II instrument generates the information in hours no days like in the case of RNA-Seq. The bioinformatics analysis is performed using the freely available SMRT® Portal software. The final result of the Iso-Seq pipeline is a list of full-length isoforms, and judging by recently published data, a significant number of the isoforms are novel [4, 52]. At this point, the Iso-Seq method is not a quantitative method but a powerful survey tool capable of identifying new isoforms.

The obvious benefits of Iso-Seq sequencing are a follows:

- Sequence full-length mRNA transcripts, with no assembly required.
- Characterize gene–isoform expression across an entire transcriptome, or within targeted regions.
- Discover novel genes and gene isoforms even in well-characterized samples.
- Perform de novo gene annotation, with or without a reference genome.
- Gather complete information about alternatively spliced exons, transcriptional start sites, polyadenylation sites, and strand orientation.
- Improve quantitation accuracy for functional genomics studies.

Even with all the benefits of using Iso-Seq, for the universal acceptance of the Iso-Seq method, the capacity of the SMRT® Cells needs to improve at least 10- to 100-fold to make the system affordable and attractive to users. The Iso-Seq method is not a drop-in replacement for the RNA-Seq method but a complementary tool. While Iso-Seq is a powerful survey tool capable of identifying new isoforms, the RNA-Seq provides all the statistical methods for measuring differential expression (DE). A hybrid approach that combines both methods is the optimal solution to study the transcriptome as demonstrated by Au's group [4].

References

1. Abdullah-Sayani A, Bueno-de-Mesquita JM, van de Vijver MJ. Technology Insight: tuning into the genetic orchestra using microarrays–limitations of DNA microarrays in clinical practice. Nat Clin Pract Oncol. 2006;3:501–16. doi:10.1038/ncponc0587.
2. Agarwal A, et al. Comparison and calibration of transcriptome data from RNA-Seq and tiling arrays. BMC Genom. 2010;11:383. doi:10.1186/1471-2164-11-383.

3. Alwine JC, Kemp DJ, Stark GR. Method for detection of specific RNAs in agarose gels by transfer to diazobenzyloxymethyl-paper and hybridization with DNA probes. Proc Natl Acad Sci USA. 1977;74:5350–4.
4. Au KF, et al. Characterization of the human ESC transcriptome by hybrid sequencing. Proc Natl Acad Sci USA. 2013;110:E4821–30. doi:10.1073/pnas.1320101110.
5. Ayub M, Bayley H. Individual RNA base recognition in immobilized oligonucleotides using a protein nanopore. Nano Lett. 2012;12:5637–43. doi:10.1021/nl3027873.
6. Bottomly D, et al. Evaluating gene expression in C57BL/6 J and DBA/2 J mouse striatum using RNA-Seq and microarrays. PLoS ONE. 2011;6:e17820. doi:10.1371/journal.pone.0017820.
7. Carneiro MO, Russ C, Ross MG, Gabriel SB, Nusbaum C, DePristo MA. Pacific biosciences sequencing technology for genotyping and variation discovery in human data. BMC Genom. 2012;13:375. doi:10.1186/1471-2164-13-375.
8. Chaisson MJ, et al. Resolving the complexity of the human genome using single-molecule sequencing. Nature. 2015;517:608–11. doi:10.1038/nature13907.
9. Chin CS, et al. The origin of the Haitian cholera outbreak strain. N Engl J Med. 2011;364:33–42. doi:10.1056/NEJMoa1012928.
10. Clontech. Manual for the SMARTer PCR cDNA Synthesis Kit. 2015. http://www.clontech.com/US/Products/cDNA_Synthesis_and_Library_Construction/cDNA_Synthesis_Kits/ibcGetAttachment.jsp?cItemId=17336&fileId=6856798&sitex=10020:22372:US.
11. Cock PJ, Fields CJ, Goto N, Heuer ML, Rice PM. The Sanger FASTQ file format for sequences with quality scores, and the Solexa/Illumina FASTQ variants. Nucleic Acids Res. 2010;38:1767–71. doi:10.1093/nar/gkp1137.
12. Dobin A, et al. STAR: ultrafast universal RNA-seq aligner. Bioinformatics. 2013;29:15–21. doi:10.1093/bioinformatics/bts635.
13. Draghici S, Khatri P, Eklund AC, Szallasi Z. Reliability and reproducibility issues in DNA microarray measurements. Trends Genet. 2006;22:101–9. doi:10.1016/j.tig.2005.12.005.
14. Eid J, et al. Real-time DNA sequencing from single polymerase molecules. Science. 2009;323:133–8. doi:10.1126/science.1162986.
15. English AC, et al. Mind the gap: upgrading genomes with Pacific Biosciences RS long-read sequencing technology. PLoS ONE. 2012;7:e47768. doi:10.1371/journal.pone.0047768.
16. Engstrom PG, et al. Systematic evaluation of spliced alignment programs for RNA-seq data. Nat Methods. 2013;10:1185–91. doi:10.1038/nmeth.2722.
17. Flusberg BA, et al. Direct detection of DNA methylation during single-molecule, real-time sequencing. Nat Methods. 2010;7:461–5. doi:10.1038/nmeth.1459.
18. Gonzalez D, Kozdon JB, McAdams HH, Shapiro L, Collier J. The functions of DNA methylation by CcrM in Caulobacter crescentus: a global approach. Nucleic Acids Res. 2014;42:3720–35. doi:10.1093/nar/gkt1352.
19. HDF_group. HDFS file format. 2015. http://www.hdfgroup.org/HDF5.
20. Kapa_Biosystems. KAPA HiFi Enzyme. 2015. http://www.kapabiosystems.com/product-applications/products/pcr-2/kapa-hifi-pcr-kits.
21. Kim D, Langmead B, Salzberg SL. HISAT: a fast spliced aligner with low memory requirements. Nat Methods. 2015;12:357–60. doi:10.1038/nmeth.3317.
22. Kim D, Pertea G, Trapnell C, Pimentel H, Kelley R, Salzberg SL. TopHat2: accurate alignment of transcriptomes in the presence of insertions, deletions and gene fusions. Genome Biol. 2013;14:R36. doi:10.1186/gb-2013-14-4-r36.
23. Korlach J, et al. Real-time DNA sequencing from single polymerase molecules. Methods Enzymol. 2010;472:431–55. doi:10.1016/S0076-6879(10)72001-2.
24. Langmead B, Salzberg SL. Fast gapped-read alignment with Bowtie 2. Nat Methods. 2012;9:357–9. doi:10.1038/nmeth.1923.
25. Langmead B, Trapnell C, Pop M, Salzberg SL. Ultrafast and memory-efficient alignment of short DNA sequences to the human genome. Genome Biol. 2009;10:R25. doi:10.1186/gb-2009-10-3-r25.

26. Larsen PA, Smith TP. Application of circular consensus sequencing and network analysis to characterize the bovine IgG repertoire. BMC Immunol. 2012;13:52. doi:10.1186/1471-2172-13-52.
27. Li JJ, Jiang CR, Brown JB, Huang H, Bickel PJ. Sparse linear modeling of next-generation mRNA sequencing (RNA-Seq) data for isoform discovery and abundance estimation. Proc Natl Acad Sci USA. 2011;108:19867–72. doi:10.1073/pnas.1113972108.
28. Life_Technologies. Manual for Trizol Plus. 2015a. https://tools.lifetechnologies.com/content/sfs/manuals/Trizol_Plus_man.pdf.
29. Life_Technologies. Manual or Poly(A)Purist™ MAG Kit. 2015b. https://tools.lifetechnologies.com/content/sfs/manuals/fm_1922.pdf.
30. Lister R, O'Malley RC, Tonti-Filippini J, Gregory BD, Berry CC, Millar AH, Ecker JR. Highly integrated single-base resolution maps of the epigenome in Arabidopsis. Cell. 2008;133:523–36. doi:10.1016/j.cell.2008.03.029.
31. Loman NJ, Quinlan AR. Poretools: a toolkit for analyzing nanopore sequence data. Bioinformatics. 2014;30:3399–401. doi:10.1093/bioinformatics/btu555.
32. Martin JA, Wang Z. Next-generation transcriptome assembly. Nat Rev Genet. 2011;12:671–82. doi:10.1038/nrg3068.
33. Mooney M, McWeeney S. Data integration and reproducibility for high-throughput transcriptomics. Int Rev Neurobiol. 2014;116:55–71. doi:10.1016/B978-0-12-801105-8.00003-5.
34. Mortazavi A, Williams BA, McCue K, Schaeffer L, Wold B. Mapping and quantifying mammalian transcriptomes by RNA-Seq. Nat Methods. 2008;5:621–8. doi:10.1038/nmeth.1226.
35. Nagalakshmi U, Wang Z, Waern K, Shou C, Raha D, Gerstein M, Snyder M. The transcriptional landscape of the yeast genome defined by RNA sequencing. Science. 2008;320:1344–9. doi:10.1126/science.1158441.
36. Nagaraj SH, Gasser RB, Ranganathan S. A hitchhiker's guide to expressed sequence tag (EST) analysis. Briefings Bioinform. 2007;8:6–21. doi:10.1093/bib/bbl015.
37. Nawy T. End-to-end RNA sequencing. Nat Methods. 2013;10(12):1144-1145 10:1144-1145.
38. Pacific_Biosciences. bas.h5 reference guide. 2015a. http://files.pacb.com/software/instrument/2.0.0/bas.h5%20Reference%20Guide.pdf.
39. Pacific_Biosciences. Metadata output guide. 2015b. http://files.pacb.com/software/instrument/2.0.0/Metadata%20Output%20Guide.pdf.
40. Pacific_Biosciences. PacBio consumables reagents. 2015c. http://www.pacificbiosciences.com/products/consumables/reagents/.
41. Pacific_Biosciences. PacBio datasets. 2015d. https://github.com/PacificBiosciences/DevNet/wiki/Datasets.
42. Pacific_Biosciences. PacBio DevNet. 2015e. http://www.pacb.com/devnet/index.html.
43. Pacific_Biosciences. PacBio SMRT Cells. 2015f. http://www.pacificbiosciences.com/products/consumables/SMRT-cells/.
44. Pacific_Biosciences. PacBio SMRT Sample Prep web site. 2015g. https://pacbio.secure.force.com/SamplePrep.
45. Pacific_Biosciences. PacBio software. 2015h. http://www.pacb.com/devnet/code.html.
46. Parkinson J, Blaxter M. Expressed sequence tags: an overview. Methods Mol Biol. 2009;533:1–12. doi:10.1007/978-1-60327-136-3_1.
47. Quick J, Quinlan AR, Loman NJ. A reference bacterial genome dataset generated on the MinION portable single-molecule nanopore sequencer. Gigascience. 2014;3:22. doi:10.1186/2047-217X-3-22.
48. Roberts A, Pimentel H, Trapnell C, Pachter L. Identification of novel transcripts in annotated genomes using RNA-Seq. Bioinformatics. 2011;27:2325–9. doi:10.1093/bioinformatics/btr355.
49. Roy NC, Altermann E, Park ZA, McNabb WC. A comparison of analog and next-generation transcriptomic tools for mammalian studies. Brief Funct Genomics. 2011;10:135–50. doi:10.1093/bfgp/elr005.
50. Sage_Science. The BluePippin System. 2015a. http://www.sagescience.com/products/bluepippin/.

51. Sage_Science. The SageELF. 2015b. http://www.sagescience.com/products/sageelf/.
52. Sharon D, Tilgner H, Grubert F, Snyder M. A single-molecule long-read survey of the human transcriptome. Nat Biotechnol. 2013;31:1009–14. doi:10.1038/nbt.2705.
53. Steijger T, et al. Assessment of transcript reconstruction methods for RNA-seq. Nat Methods. 2013;10:1177–84. doi:10.1038/nmeth.2714.
54. Steinbock LJ, Radenovic A. The emergence of nanopores in next-generation sequencing. Nanotechnology. 2015;26:074003. doi:10.1088/0957-4484/26/7/074003.
55. Trapnell C, Pachter L, Salzberg SL. TopHat: discovering splice junctions with RNA-Seq. Bioinformatics. 2009;25:1105–11. doi:10.1093/bioinformatics/btp120.
56. Trapnell C, et al. Transcript assembly and quantification by RNA-Seq reveals unannotated transcripts and isoform switching during cell differentiation. Nat Biotechnol. 2010;28:511–5. doi:10.1038/nbt.1621.
57. Walter NA, McWeeney SK, Peters ST, Belknap JK, Hitzemann R, Buck KJ. SNPs matter: impact on detection of differential expression. Nat Methods. 2007;4:679–80. doi:10.1038/nmeth0907-679.
58. Wang K, et al. MapSplice: accurate mapping of RNA-seq reads for splice junction discovery. Nucleic Acids Res. 2010;38:e178. doi:10.1093/nar/gkq622.
59. Wang L, Feng Z, Wang X, Wang X, Zhang X. DEGseq: an R package for identifying differentially expressed genes from RNA-seq data. Bioinformatics. 2010;26:136–8. doi:10.1093/bioinformatics/btp612.
60. Wang L, Si Y, Dedow LK, Shao Y, Liu P, Brutnell TP. A low-cost library construction protocol and data analysis pipeline for Illumina-based strand-specific multiplex RNA-seq. PLoS One. 2011;6:e26426. doi:10.1371/journal.pone.0026426.
61. Weber AP, Weber KL, Carr K, Wilkerson C, Ohlrogge JB. Sampling the Arabidopsis transcriptome with massively parallel pyrosequencing. Plant Physiol. 2007;144:32–42. doi:10.1104/pp.107.096677.
62. Wilhelm BT, et al. Dynamic repertoire of a eukaryotic transcriptome surveyed at single-nucleotide resolution. Nature. 2008;453:1239–43. doi:10.1038/nature07002.
63. Wu TD, Nacu S. Fast and SNP-tolerant detection of complex variants and splicing in short reads. Bioinformatics. 2010;26:873–81. doi:10.1093/bioinformatics/btq057.
64. Zerbino DR, Birney E. Velvet: algorithms for de novo short read assembly using de Bruijn graphs. Genome Res. 2008;18:821–9. doi:10.1101/gr.074492.107.
65. Zheng CL, Kawane S, Bottomly D, Wilmot B. Analysis considerations for utilizing RNA-Seq to characterize the brain transcriptome. Int Rev Neurobiol. 2014;116:21–54. doi:10.1016/B978-0-12-801105-8.00002 3.

Author Biography

Dr. Manuel L. Gonzalez-Garay is an assistant professor at the University of Texas at Houston (UTHealth). He is a faculty member of the Center for Molecular Imaging at the Institute of Molecular Medicine and director of the Division of Genomics and Bioinformatics at the Center for Molecular Imaging. He is also an affiliated professor at the Universidad Autonoma de Nuevo Leon, Mexico. Dr. Gonzalez-Garay has been working as a bioinformatician for the last 19 years and has extensive experience in genetics, cell biology, molecular biology, bioinformatics, software development, and computational analysis. Dr. Gonzalez-Garay has been working as a bioinformatician at pharmaceutical companies and also at the

Baylor College of Medicine Human Genome Sequencing Center (BCM-HGSC). At BCM-HGSC, he led the development of bioinformatics software and was involved in multiple resequencing projects. He also participated in a significant number of sequencing projects including the Tumor Sequencing Project (TSP), The Cancer Genome Atlas (TCGA), and the sequencing and assembly of various organisms including the rat genome. He is currently working on the identification of genes involved in rare disorders with a particular emphasis on lymphatic disorders. He has published over 25 scientific papers in internationally recognized journals. His papers have been cited over 6800 times according to Google Scholar, and he has an h-index of 15.

Chapter 7
Systematic and Integrative Analysis of Gene Expression to Identify Feature Genes Underlying Human Diseases

Zixing Wang, Wenlong Xu and Yin Liu

Abstract Over the past two decades, the advances in genomics technology have opened the door for rapid biological data acquisition and have revolutionized many aspects of biomedical research. Given the complex and noisy nature of the large-scale biological data, there is a high demand for developing variable selection approaches to identifying disease biomarkers in the field of translational bioinformatics. These biomarkers offer early detection of pathogenesis, inform prognosis, provide guidance for the treatment, and monitor disease progresses. In this chapter, we focused on developing a variety of methods that systematically analyzed whole-genome gene expression data for identifying feature genes associated with patient clinical parameters. In the first method, we constructed a gene co-expression network and then selected genes that are informative for classifying different cancer subtypes based on gene connectivity within the co-expression network. In the second method, we incorporated prior biological pathway information to reconstruct a gene network and then identified hub genes that are associated with cancer prognosis. Finally, we identified protein subnetworks instead of individual genes as biomarkers for classifying different types of brain injuries. Our study has set up a framework that can be easily generalized to integrate different types of genomics and proteomics information for better identifying feature genes to improve accuracy of disease diagnosis and treatment.

Z. Wang
Department of Bioinformatics and Computational Biology, University of Texas MD Anderson Cancer Center, 77455 Fannin Street, Houston, TX, USA

W. Xu · Y. Liu (✉)
Department of Neurobiology and Anatomy, University of Texas Health Science Center at Houston, Medical School Building, Suite 7.046, 6431 Fannin Street, Houston, TX 77030, USA
e-mail: Yin.Liu@uth.tmc.edu

Y. Liu
University of Texas Graduate School of Biomedical Science, 6767 Bertner Avenue, Houston, TX, USA

J. Wu, *Transcriptomics and Gene Regulation*,
Translational Bioinformatics 9, DOI 10.1007/978-94-017-7450-5_7

Keywords Gene expression biomarkers · Feature selection · Protein interaction · Network

7.1 Introduction

Within the past decade, there has been a growing demand for proper usage of disease biomarkers in clinical diagnosis and prognosis. Disease biomarkers are typically selected according to their power in discriminating different disease states. They offer early detection of pathogenesis, inform prognosis, provide guidance for the treatment, and monitor disease progresses. In the functional genomic era, recent advances in high-throughput technologies open up for new opportunities that can meet challenges in the area of biomarker identification. For example, when the expression profiles of thousands of genes are available, the biomarker discovery can be modeled as a feature gene selection problem that tends to find the most relevant features (genes) for appropriate disease classification [1].

Figure 7.1 shows the development of a gene expression biomarker from large-scale gene expression data for cancer prognosis. Cancer cells often involve multiple genomic changes that together are responsible for uncontrolled cell growth. As cellular behavior is controlled by gene activity, it is logical to assume that differences in the tumor samples could be inferred from differences in their gene expression. Therefore, gene expression profiles can be used on the identification of prognostic biomarkers in cancer. Given the gene expression profile on tumor samples with known clinical outcome, the set of genes that correlates best with the relevant clinical parameters can be identified by feature selection methods. Next, the gene signature is validated on another set of independent samples of known clinical outcome, and its performance is evaluated. Then, the gene signature needs to go through regulatory approval before it is used in clinical setting. Finally, patients prognosis outcome can be predicted and classified based on the gene signature.

Feature selection has been well studied in the context of supervised learning where the label information is available [2, 3]. It evaluates the relevance of features by the extent of alignment between features and the class label. However, in practice, there are usually abundant features but lack of class label information. Under such circumstances, unsupervised feature selection becomes an alternate solution. One of the application domains of unsupervised feature selection is on selecting relevant feature genes for clustering similar samples into one group in an unsupervised manner. In this context, feature selection algorithms can be categorized as either the wrapper or the filter approaches according to the evaluation criterion in searching for relevant features. For the wrapper approaches, each candidate feature or feature subset is obtained by conducting a combinatorial search in full feature space and each candidate is evaluated by measuring its goodness with a specific clustering algorithm [4, 5]. The wrapper approaches have shown their

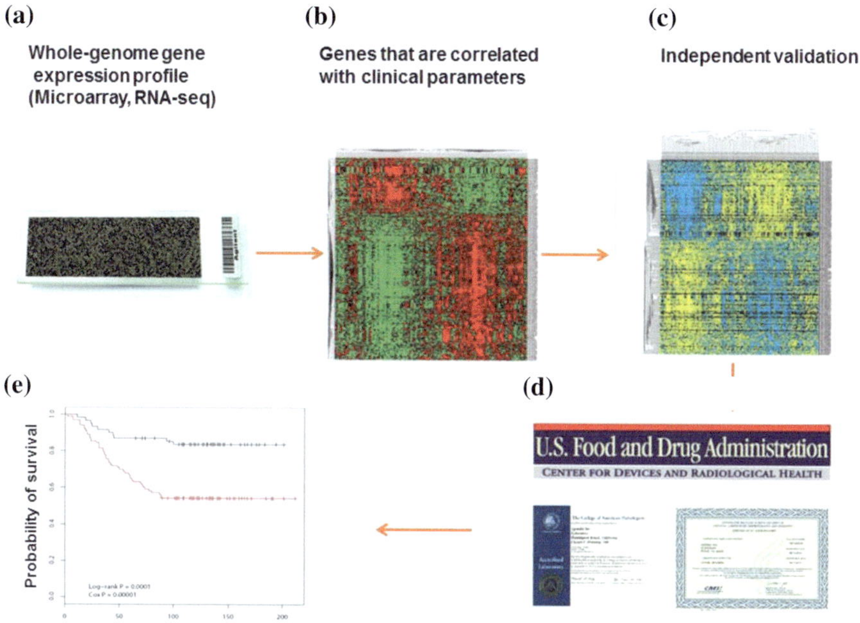

Fig. 7.1 The steps for developing a gene expression biomarker. **a** The whole-genome gene expression profile on tumor samples with known clinical outcome. **b** The set of feature genes correlated with clinical parameters can be identified by bioinformatics approaches. **c** The gene signature is evaluated and validated with an independent set of samples with known clinical outcome. **d** The identified gene signature is subject to regulatory approval. **e** The approved gene signature can be used to stratify patients

effectiveness on low dimensional data. However, one problem of these methods, when applied to large datasets, is the high computational complexity since the size of candidate subset space is exponentially increased with the number of features. Furthermore, the wrapper approach lacks robustness and is biased toward the specific clustering algorithm used [6]. In contrast, the filter approach is more efficient in that it does not utilize any clustering algorithms to evaluate the candidate features. Instead, features are evaluated according to certain criteria (e.g., feature variance, entropy-based distance [7], similarity among feature [8], and Laplacian score [9]), and then, a number of less informative features are filtered out before clustering algorithm is performed. This approach is much faster and more efficient than the wrapping method in high-dimensional dataset analysis.

Despite the wide application of these approaches on feature selection problems, there has relatively limited success in applying these techniques to better clarify and characterize the clinical heterogeneity observed for many complex diseases. In recent years, high-throughput gene expression profile measured by microarray and next-generation sequencing (NGS) techniques has been proven as an informative platform by which genome-scale events can be translated into medical practice.

For example, based on gene expression profiles, interesting cluster distinctions can be identified among patients, which may correspond to particular phenotypes, such as different clinical syndromes or cancer types [10]. Under such circumstances, many traditional feature selection methods fell short to capture the full spectrum of the underlying structure of sample clusters. It is noted that gene expression could be influenced by the cell type, cell phase, external signals, and many more factors; therefore, the gene expression dataset can be contributed by all these factors mixing together. Multiple meaningful clustering structures can coexist based on the same dataset, and the same set of samples can undergo different partitions based on different subsets of selected feature genes. Thus, a good feature selection algorithm should be able to select informative genes that best preserve multiple clustering structures in the data.

In this chapter, we describe novel methods to achieve feature gene selection for downstream clustering/classification analysis [11–14]. These methods are based on systematic and integrative analysis of gene expression profiles. In the first method, we conceptually build a gene correlation, or co-expression network, in which nodes and edges represent genes and their expression similarity, respectively. This network paradigm is based upon the assumption that each gene may induce a specific partition of the sample space in the absence of a priori information about the variable space. Therefore, given the thousands of genes in the high-throughput expression profile, there might be a number of distinct sample clustering solutions on the same set of samples. In this context, the imminent task is to combine these multiple partitions into a single consensus clustering, which should share as much information as possible with the given pool of sample partitions. This notion of integrating multiple clustering solutions is in line with the framework of cluster ensembles [15], which tend to reuse the existing knowledge and minimize the information loss incurred in the process of cluster assembling. Based on the assumption that higher correlated gene expression profiles tend to produce more similar partition structures, we propose to assemble genes according to their expression similarity rather than their sample partitions. Following this line of reasoning, we select a list of individual genes that sustains the most similarity with other genes, so that the final sample partition based on this gene list is a combination with the most consensus information among the partitions inferred by each individual gene.

We note that a gene network constructed exclusively from expression information will neglect prior biological knowledge or information about the gene interaction. As a result, one-dimensional aspect of gene expression analysis may overlook the intrinsic relationships among genes. Recently, integrative analysis of gene expression has received a great deal of attention. A multi-dimensional characterization of the genomic data has become a standard practice. This is particularly true when a vast repository of prior biological knowledge has been rapidly accumulated over past few years. There is a strong interest in leveraging this prior information to effectively interrogate the genomic expression data and to achieve the goal of identifying genes that may jointly influence a biological response. For this purpose, we have developed a L1 penalized least square estimator, named the prior lesso (pLasso) method, for the reconstruction of gene networks [13]. In this method, we partition edges between genes into two subsets: One subset of edges is present in known pathways, whereas the other has no prior information associated. Our method assigns different prior distributions to each subset according to a modified Bayesian information criterion that incorporates prior knowledge on both the network structure and the pathway information. After the gene network is reconstructed, the top genes with the most neighbors in the inferred network were selected as feature genes that are subject to further testing on their power in predicting survival outcome of cancer patients. Finally, in the third method we will describe in the chapter, we aim to identify biomarkers not as individual genes but as gene subnetworks. In traditional expression profiling studies, genes that are not significantly differentially expressed between classes are often neglected. These discarded genes' modest association with specific phenotypes may represent false negatives and may be important biomarkers. We expect that these genes may be identified within functional unites of genes that in aggregate have a significant association to a specific phenotype. Therefore, we develop a network-based approach incorporating gene expression profiles and the protein–protein interaction information from existing databases to identify gene subnetwork biomarkers and apply this approach on traumatic brain injury (TBI) study for characterizing classes of TBI.

7.2 Leveraging Gene Co-expression Networks in Feature Selection

In line with the framework of ensemble clustering, the connectivity of one gene node in the co-expression network is used to estimate the information gain as a filter criterion throughout the network. The gene connectivity is defined as the degree of similarity between its expression profiles with others. We design a simple network inference method to construct the co-expression network in Sect. 7.2.1 and then propose a transformed gene connectivity measure for module recovery in Sect. 7.2.2. Sections 7.2.3 and 7.2.4 show the performance of our method based on simulation datasets and real dataset analysis, respectively.

7.2.1 Gene Co-expression Network Analysis and Gene Connectivity

We define the similarity s_{ij} between the expression profiles of genes i and j using the absolute value of the Pearson correlation $s_{ij} = \text{abs}(\text{cor}(x_i, x_j))$, where x_i and x_j represent the gene expression profiles for genes i and j, respectively. Therefore, the similarity matrix can be denoted by $S_{ij} = [s_{ij}]$. We then transform this similarity matrix into an adjacency matrix using a "soft" power transformation function [16]:

$$a_{ij} = \left|s_{ij}\right|^{\beta} \tag{7.1}$$

with a single parameter β, where $\beta \geq 1$. Here, a_{ij} is an element of the adjacency matrix. Soft-thresholding results in a completely connected network with each edge being assigned a weight.

Given a $n \times n$ symmetric adjacency matrix, the connectivity k_i of gene i is defined by

$$k_i = \sum_{j \neq i} a_{ij} \tag{7.2}$$

For the soft-thresholding transformation, the connectivity of gene i equals the sum of weights between gene i and all other genes in the weighted network. To select the relevant genes for clustering analysis, we first rank the genes according to their connectivity, which can be obtained for all the genes. The genes with low ranks are filtered, while the genes with top ranks are considered to have high degree of connectivity and are selected for clustering analysis.

7.2.2 Module Identification and Visualization

A gene co-expression network can be used to identify co-regulated subsets of genes, known as modules, such that genes within a module are more highly correlated than those between modules. Usually, a module corresponds to certain function unit in the complex network, such as different biological processes or pathways. We introduce a new module identification and visualization method. Our module identification method is based on using a node similarity measure of their relative interconnectedness coupled with the hierarchical clustering method. We calculate the Jaccard similarity coefficient J_{ij} to represent the similarity measure.

$$J_{ij} = \frac{h_{ij}}{k_i + k_j - h_{ij}} \tag{7.3}$$

where $h_{ij} = \sum_u a_{iu}a_{uj}$, which equals the total interconnectedness of genes i and j in the soft-thresholding transformation. Therefore, the similarity measure will be affected by the selection of the transformation parameters. In our implementation, we adjust the power function parameter β to explore their effects on the results of variable selection. Once the similarity measure matrix is obtained, we reorder it by hierarchical clustering of each row and column to put similar genes in an adjacency zone [17]. Since the similarity measure matrix is symmetric, these highly similar genes would form a "hot" block along the diagonal in the heatmap and can be identified as a module by visual inspection. The genes in the resulting modules are expected to be tightly connected to each other in the gene co-expression network and thus are highly co-expressed.

7.2.3 Feature Selection Performance in Simulation Studies

We used a simulation setting similar to that in Witten and Tibshirani [18]. Each simulation dataset contains 60 samples from three classes C_1, C_2, and C_3 (20 samples from each), and each sample X_i is a d—dimensional vector that follows $N(\mu_i, \Sigma_d)$ and is independent of other samples. Thus, the clustering structure is determined by the specification of μ_is that are defined as

$$\mu_{ij} = \begin{cases} \mu(1_{i \in C1} - 1_{i \in C \setminus 1}) & \text{if } j \leq 10 \\ \mu(1_{i \in C \setminus 1} - 1_{i \in C1}) & \text{if } 10 < j \leq 20 \\ \mu(1_{i \in C2} - 1_{i \in C \setminus 2}) & \text{if } 20 < j \leq 30 \\ \mu(1_{i \in C \setminus 2} - 1_{i \in C2}) & \text{if } 30 < j \leq 40 \\ 0 & \text{otherwise} \end{cases} \tag{7.4}$$

where μ is a positive constant and set to 1 in the experiment. This configuration of μ sets the first 40 genes as informative genes, designating the other genes as noise. We take $\Sigma_d = \text{diag}(\sigma_1, \ldots, \sigma_d)$ where $\sigma_1, \ldots, \sigma_d$ are set such that the population variance of each variable is one. In the simulation, the first 20 genes together can be considered as a module since their expression profiles are highly correlated and this module differentiates samples in class C_1 from the others, whereas the next 20 genes form another module that differentiates samples in class C_2 from others. Therefore, these two sets of genes exhibit different sample partitions.

Given the simulation dataset, we evaluate the performance of our variable selection method based on the soft-thresholding transformation. Herein, we use F-score and classification error rate (CER) as measuring metrics, where $F = 2 \times \text{Precision} \times \text{Recall}/(\text{Precision} + \text{Recall})$. For CER, it is defined as the deviation of sample clustering (p_1) from the true clustering labels (p_2) with following formula:

$$\text{CER}(p_1, p_2) = \frac{\sum_{i > i'} \left| 1_{p_1(i) == p_1(i')} - 1_{p_2(i) == p_2(i')} \right|}{\binom{n}{2}} \tag{7.5}$$

where n is the sample size. Note that smaller CER values reflect more accurate clustering results. A CER of zero indicates that the clustering results p_1 and p_2 agree perfectly.

The soft-thresholding transformation is only dependent on the power function parameter β. As shown in Fig. 7.2, the power transformation significantly improved

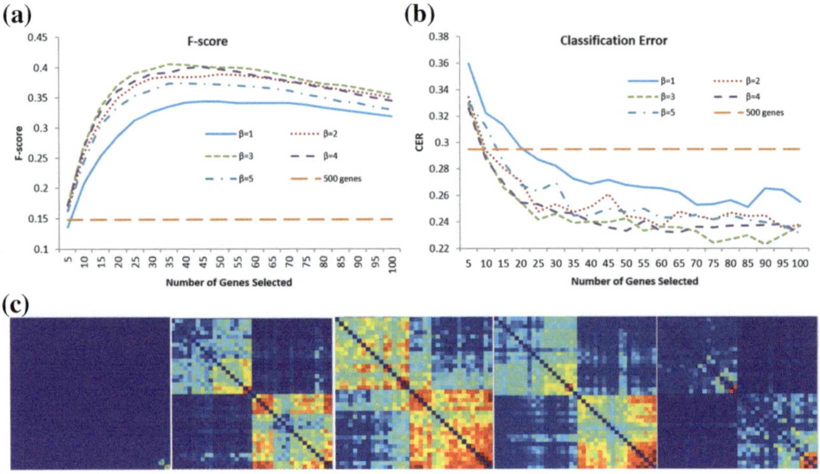

Fig. 7.2 Performance of feature gene selection from the simulated dataset with 500 genes. **a** The average of F-scores and **b** the classification error curves based on soft-thresholding transformation with different values of power function parameter β. The *horizontal line* in **a** and **b** represents the performance based on all genes without the feature selection step (500 genes totally). **c** The first *block shows* the module structure in the gene co-expression network. The rest on the *right* are zoomed-in view of the modules highlighting the genes included in two modules with the power function parameter β = 2, 3, 5 and 7, respectively

the performance of variable selection and led to a higher F-score peak and lower CER than the original non-transformed one (β = 1). We further found the performance was not a monotonic function of β. Among the four power functions with different parameters β, the optimal value of F-score and CER was achieved when β was set to 3. This indicates that 3 is the optimal power function parameter in the simulated dataset, and it results in an optimized state for emphasizing the correlation associated with true gene relationships by diminishing the noisy effects.

We also have applied the soft-thresholding transformation on the gene co-expression network for module identification. Note that the sensitivity of this method varies depending on the co-expression network size and the composition of variable space. In the analysis, we assigned the value of u to 1.5 in the simulation setting to demonstrate a clear module structure. Figure 7.2c shows the discovered modules in the network when the power function parameters are varied in the transformed network. Two "hot blocks" can be clearly identified along the diagonal, each of which corresponds to the original defined module in the simulation setting with a few missing genes. Although due to varying values of the transformation parameters, the boundaries between blocks exhibited distinctive sharpness (Fig. 7.2c), the module structure, and the genes included in each module were the same, indicating the relative robustness of our method in module identification. We also evaluated their clustering performance. Each of the blocks induced a specific bipartitioning of the sample space that is equivalent to the sample partition inferred by the corresponding modules in our simulation setting.

7.2.4 Real Data Application

Along with simulations, we have applied our method to two real experimental datasets: leukemia [19] and colon cancer data [20]. The leukemia dataset consists of 72 patients with two subtypes of acute leukemia: acute myeloid leukemia (AML) and acute lymphoblastic leukemia (ALL). The latter is composed of two subclasses, B-cell and T-cell types. Therefore, there could be two possible biologically meaningful clustering solutions including one with two clusters of samples (AML and ALL) and the other with three clusters (T-cell ALL, B-cell ALL, and AML).

Following Dudoit and Fridlyand [21], three preprocessing steps are applied to the original data matrix and a final 72 × 3571 data matrix was obtained. Because the preprocessing steps included thresholding the gene expression values with a floor and a ceiling boundaries, many artificially high correlations can be introduced. We filter out these genes whose medians equal to the boundary values and obtained 3033 genes in total. We first studied the module organization of the gene space and the associated sample partition in the leukemia dataset. As shown in the Fig. 7.3a, the topological similarity matrix exhibited a sharp separation of modules from its

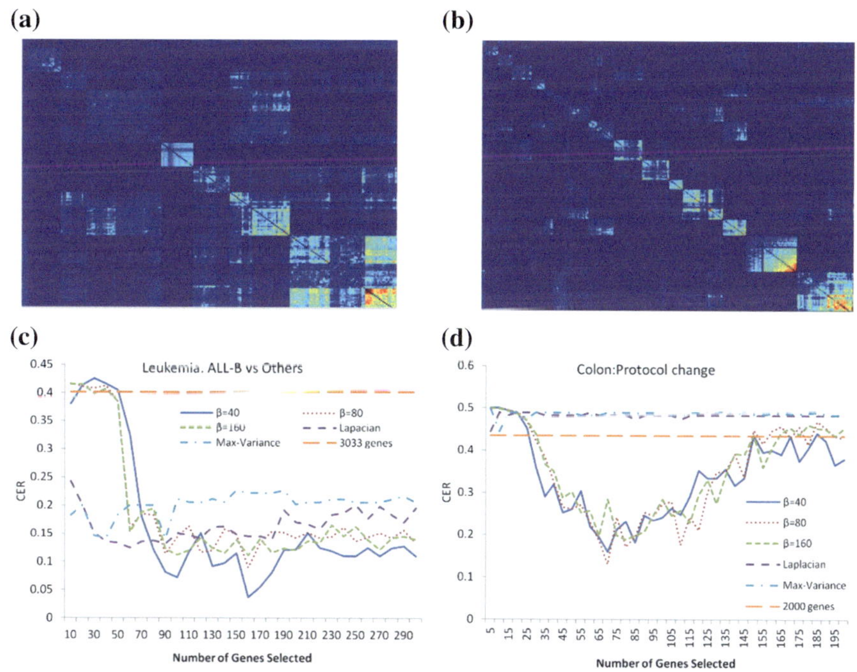

Fig. 7.3 Module analysis and clustering results for the leukemia and colon datasets. *Top panel* Zoomed-in view of the module structure in the gene co-expression network of the leukemia (**a**) and colon dataset (**b**). *Bottom panel* The clustering performance for new partition structures based on soft-thresholding transformation with various power functions, Laplacian score method, or Max-variance method. **c** Leukemia dataset and **d** colon cancer dataset

neighboring genes. We evaluated the sample clustering performance of modules by using the gene set included in each module for sample partitioning. We found that most of them induced a meaningful partition of the sample space. Specifically, the first module at the bottom right corner rendered a dichotomy of the samples according to the known classification, ALL/AML, with the CER value equaling 0.155, whereas the second module tends to distinguish B-cell ALL patients from the rest with a CER value of 0.2, indicating the unrecognized similarity between ALL T-cell samples and AML samples in the dataset. The other modules also impose a potential novel partition of samples. These results confirmed multiple possible clustering solutions in the leukemia dataset. We also performed feature selection to select individual genes based on their connectivity in the transformed network using soft-thresholding transformation. For the three-class solution, the 100 genes selected from the network-based analysis yielded the sample partition coinciding almost precisely with the known classification (T-cell ALL, B-cell ALL and AML) with CER equaling to 0.08. The genes selected from our method may also represent new sample partition. This is supported by the observation that our method achieved better separation of B-cell AML samples from the rest, compared with other filter methods. Results in Fig. 7.3c showed that our approach achieved a better optimal CER value compared with the Laplacian score and the maximum variance methods, with a comparable number of genes selected.

We also analyzed the colon cancer data [20]. The colon dataset contains two classes of samples based on disease status with 62 total samples: 40 tumor samples and 22 normal samples. However, an independent study reported that there was an inconsistency in the experimental protocols used to process the dataset [22]. The 22 samples including 11 tumor and 11 normal samples were processed by protocol 1, and the rest of 40 samples were processed by protocol 2. Taking the different protocols into consideration, the study has at least three different possible sample partition structures based on disease status, sample protocols, and their combination. Dozens of modules were identifiable along the diagonal of the similarity matrix (Fig. 7.3b). Each module exhibits a distinctive partition of the sample space. Among the first three modules at the bottom right, module 1 had strong tendency to partition the samples according to the normal versus tumor classification with a CER value (0.35), whereas module 3 was informative for the partition based on different protocols (CER = 0.27). Also, the number of genes included in these two modules differed. Module 1 was the biggest among these 3 modules in terms of the number of genes included. These results indicate that the classification of tumor versus normal samples is a more dominant factor in the sample clustering compared to the different sample protocols. It was interesting to observe that the number of genes in module 2 was similar compared to module 1. However, their clustering behaviors differed, suggesting that module 2 may inform a novel sample partition of the colon cancer dataset. The aforementioned module analysis reinforced the idea that the colon cancer dataset has at least three different clustering solutions. In a new two-class solution where the samples were processed by two different protocols, our approach achieved a much better performance than the other methods with a different number of genes selected (Fig. 7.3d). For instance, when β was 40 or above, our method had a CER value equal to 0.15 with 70 genes selected, whereas the Laplacian score and maximum variance methods fared much worse.

We further examined the selected gene lists with varied power transformation parameter β in both datasets. We found that the overlap among the gene lists increased as β became larger. After β reached 80 for the leukemia dataset, the gene lists essentially remained unchanged. This is consistent with their similar clustering performance as shown in Fig. 7.3c, d. However, in the simulation studies, the optimal value of β was small ($\beta = 3$), reflecting different variable compositions in these two datasets. As shown in Fig. 7.4, these implications became clear that, unlike 40 informative genes in simulation (Fig. 7.4a), the correlation of the 100 selected genes has a mixture distribution with two components, one at the high end

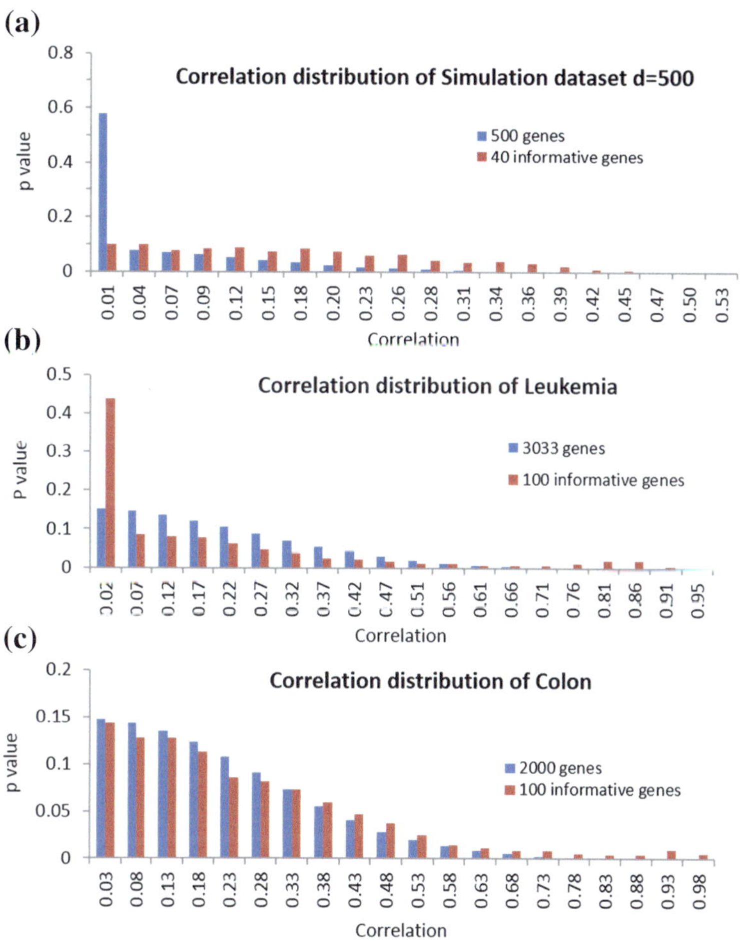

Fig. 7.4 The correlation distribution of full variable space and informative gene set. **a** The simulated dataset with 500 genes, **b** the leukemia, and **c** the colon cancer dataset

of the distribution and another is close to zero (Fig. 7.4b). Since the component in the high end was well separable from the other for the full gene space, the power function will aggravate the situation so that these highly connected genes tend to be selected as informative genes disregarding the value of β parameter. This may explain the saturating effect of β parameter in the performance. Given the assumption that experimental gene expression data represent several active biological processes, where genes corresponding to each of the process tend to be highly correlated with one another but not well correlated with genes participating in other biological processes, it is reasonable to expect that in real expression analyses, the gene sets will have a mixture distribution with multiple well-separated components for their pairwise correlation. Therefore, the selected genes should not be sensitive to the power function parameter β when it is above a certain threshold. This was further validated in the colon cancer data analysis (Fig. 7.4c). Similar to the leukemia dataset, the gene set has two well-separated clusters of correlation at two ends, which may explain their saturating behavior of clustering performance when β reaches a certain value. In conclusion, both simulation studies and real data application have demonstrated that our feature selection method has better performance in terms of the reliability of the selected genes and sample clustering results. In addition, a module recovery method is developed that can help discover novel sample partitions that might be hidden when performing clustering analyses using all available genes.

7.3 Incorporating Prior Pathway Information to Identify Genes Associated with Cancer Prognosis

We propose a gene network reconstruction method incorporating prior biological knowledge on gene pathways or gene–gene interactions. Based on the inferred gene network, we select the highly connected genes aiming to accurately discriminate different disease types. Instead of focusing on the marginal correlation described in the previous section, here we perform partial correlation analysis to construct the gene network through a least square regression approach. The partial correlation measures the direct association between two genes in the gene association network. Therefore, compared to the marginal correlation, it has an important advantage in the network inference where it enables to distinguish direct from indirect gene associations that originate via intermediate genes (sequential pathways) or due to other genes (common causes). With this adaption, we can easily derive partial correlation from a least square regression approach, which creates additional capacity to incorporate prior biological knowledge on gene networks. In Sect. 7.3.1, we describe the rationale of the prior-dependent lasso-based network reconstruction method, named pLasso. We implement this method on both simulated and real datasets and then evaluate the performance of highly connected genes from the inferred network based on its classification power in Sects. 7.3.2 and 7.3.3.

7.3.1 Lasso-Based Network Reconstruction

The gene network can be described by a graph $G = (V, E)$, where V is the node set representing genes and E is the edge set representing conditional independence. We can model the joint distribution of multiple genes with the Gaussian graphical model (GGM). In the GGM, we assume the available data X is a random variable generated from a multivariate Gaussian distribution with mean zero and covariance matrix. X can be obtained from large-scale genomic information, such as the microarray or the RNA-seq data. The inverse of the covariance matrix, known as the precision matrix, describes the conditional independence structure of X. The precision matrix can be easily linked to the partial correlation in the graphical model, where the pattern of zero entries in the matrix corresponds to conditional independence between variables. However, the genomic data present modeling challenges due to the small n (the number of samples) but large p (the number of variables) problem. To overcome this problem, we designate lasso-based regularized high-dimensional regression [23–25]. In this study, we use Meinshausen and Buhlmann's [26] neighborhood selection method. Basically, lasso regression is applied to each node in the network, reducing the original problem to multiple sparse linear regression problems.

Given the regression coefficients β, where each gene is considered as the response variable sequentially and all the other genes are the covariates, the partial correlation coefficients can be derived. To incorporate the prior information in network inference, we present a Bayesian interpretation for lasso regression. Tibshirani [27] indicated that the lasso estimate can be viewed as the model of the posterior distribution of β with a double exponential distributed prior (or Laplacian prior) by maximizing the log posterior distribution of

$$p\left(\widehat{\beta}|X\right) \sim C\exp\left\{-\frac{1}{2}\left(\left\|X^{(i)} - X^{(\backslash i)}\beta\right\|^2 + \lambda\sum_j |\beta_j|\right)\right\} \tag{7.6}$$

where C is a constant. Thus, the lasso penalty can be regarded as the logarithm of the prior distribution of the parameter $\beta = (\beta_1, \ldots, \beta_p)^T$, which is a Laplacian prior with mean equal to 0. Because prior distributions model our prior knowledge of the data, the known network structure can be introduced in a very natural way in the form of prior probabilities. A mixture of two Laplacian prior distributions for the regression coefficients is proposed as in Eq. (7.7) with different parameters λ_1 and λ_2.

$$p(\beta|\lambda_1, \lambda_2) \sim \exp\left\{-\lambda_1\sum\left|\beta_{\text{non-prior}}\right| - \lambda_2\sum\left|\beta_{\text{prior}}\right|\right\} \tag{7.7}$$

Here, λ_1 and λ_2 are regularization parameters. $\beta_{\text{non-prior}}$ and β_{prior} represent the regression coefficients corresponding to the edges absent and present in the prior knowledge. The prior distribution of regression coefficients for the edges not present in known databases is concentrated and then enforces most regression coefficients shrinking to 0. On the other hand, the prior distribution of regression coefficients corresponding to existing edges in the known databases is diffuse. A reliable data source reflects increased confidence in the gene interaction. Their regression coefficient profile is scattered away from zero. Therefore, it is preferable to include the regression coefficients representing the known gene interactions in the lasso-based modeling. In our pLasso method, we select different values of the regularized parameter λ (λ_1 and λ_2) in two lasso penalty terms, thus allowing the lasso regression coefficients corresponding to the edges absent and present in the prior knowledge to have different prior distributions. However, how to assign these two regularization parameter is a challenging problem. In this study, we propose a modified BIC score for regularization parameter selection [13].

7.3.2 Simulation Studies

To demonstrate the performance of the proposed pLasso method, we conducted simulation studies to empirically compare our method with the traditional lasso method. In the experiment, we designed two simulation scenarios on different network scales. The first simulated network is a small network with 40 nodes, an average node degree of 4 and a maximum degree of 6. The 80 ($40 \times 4/2$) edges were randomly assigned to the 780 ($40 \times 39/2$) node pairs with the limit that the maximum node degree is not exceeded. According to this network structure, we simulated its associated gene expression datasets similar to others [24, 25]. Basically, we first constructed a positive definite partial correlation matrix P based on the simulated network. Then, the gene expression data were simulated from a standard multivariate normal distribution with correlation structure derived from P with each gene having 10, 100, and 200 replicates so that we could further investigate the effect of sample size on the performance of our method. Our second simulation scenario has a larger network with 300 nodes and 900 edges, an average degree of 6 and a maximum degree of 12. Each gene was simulated with 10, 100, and 200 replicates. In real situations, the true underlying network is only partially known and is mixed with spurious edges. To investigate the performance of our pLasso method with imperfect prior information, we have simulated prior information with different precision levels varying from 0.1 to 1.0. The total number of edges in the prior data was set equal to the number of edges in the true underlying network. Therefore, a precision level of 0.1 indicates that 10 % of the edges in prior are true edges, while the other 90 % are spurious ones, and a precision level of 1.0 indicates a perfect prior with all true edges included. To incorporate the prior information, the network recovery method pLasso was implemented to search over a sequence of μ from 0 to 1.2 with an increment of 0.1 to get the optimal λ_2, where $\lambda_2 = \mu \times \lambda_1$. The optimal λ_2 value was obtained when the minimum BIC score was achieved. The parameter λ_1 used in pLasso was set the same as that in the original lasso approach based on the BIC criterion. We have demonstrated that in all simulation scenarios, combining prior knowledge with higher precision in pLasso led to a higher F-score, a weighted average of the precision and recall rates of the method. Nevertheless, we found F-scores from pLasso were consistently higher than those from a traditional lasso method. Even when the precision of the prior was as low as 0.1, pLasso achieved an F-score comparable to or slightly higher than did the lasso method for all the simulation scenarios [13]. Therefore, even if perfect prior information cannot be obtained in practice, our approach helps to distinguish the true edges from the spurious ones and outperforms a traditional lasso method that neglects prior information.

7.3.3 Integrative Analysis of Breast Cancer Gene Expression Data

To evaluate the performance of the pLasso method, we applied it to a publicly available dataset for network inference. This is a microarray gene expression study of breast cancer [28], measuring gene expression profiles of 286 lymph-node-negative breast cancer patients. Among these patients, 107 patients have developed a distant metastasis, whereas 179 patients are metastasis free. In the analysis of breast cancer data, our objective is to reconstruct the gene regulatory networks for two pathophenotypes of breast cancer, the metastasis-free group and the group with metastases. Since the performance of lasso and pLasso are sensitive to the sample size, we only used 107 of 179 metastasis-free patients so that the sample size of the metastasis-free group is the same as that of the patients with metastases. We utilized a prior gene network compiled from the KEGG database and the Pathway Commons Web resource. With traditional lasso method, the inferred network from patients with metastases had 21,360 edges. For the pLasso setting, we set the precision of the prior knowledge to 0.6, as we expect that 60–70 % of the edges present in the prior knowledge would align with the interactions in the true network corresponding to breast cancer samples [29]. Both patient groups resulted in networks with similar number of edges. In the groups of patients with metastases, we inferred a network with 5187 genes and 29,821 edges, whereas the metastasis-free patient group yielded a network with 5106 genes and 29,364 edges. An example showing the difference between the inferred networks from the lasso and pLasso approaches in patients with metastases is illustrated in Fig. 7.5. The metastatic progression of breast cancer is directly caused by the disregulation of numerous cellular signaling pathways. BUB1B, as a protein kinase, has been known to be essential in the mitotic checkpoint during normal mitosis progression. Recently, an analysis on multiple public datasets of gene expression discovered that BUB1B is associated with early distant metastases in breast cancer [30]. Here, we took BUB1B and its neighbors to exemplify the inferred network structure difference in breast cancer patients with metastases. As shown in Fig. 7.5, in the group of breast cancer patients with metastases, BUB1B possessed a higher node degree of

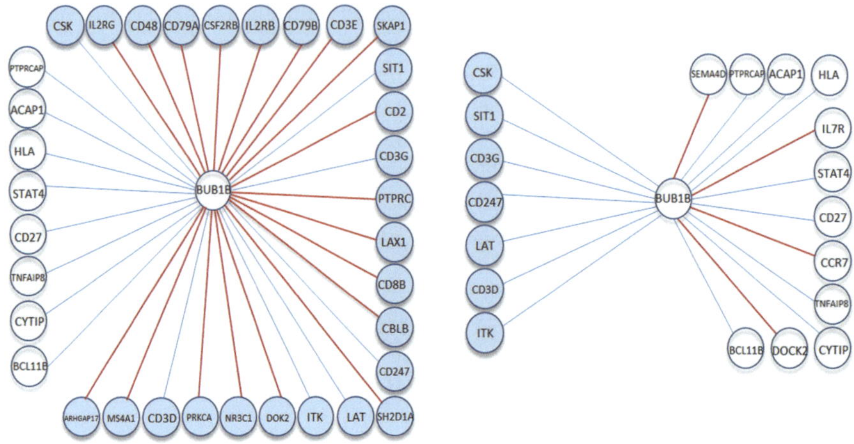

Fig. 7.5 BUB1B and its neighbor genes inferred from pLasso (*left*) and lasso (*right*) methods in patients with metastases. The *circles shaded in blue* indicate the inferred neighbor genes existed in the prior databases. The edges in *red mark* the difference between lasso and pLasso inferred networks

34 in the pLasso inferred network than that in the lasso inferred network (node degree of 19). We found 26 out of 42 edges present in the prior knowledge were included in the pLasso inferred network, while only 7 edges in the prior were recovered in the Lasso inferred network. As expected, one effect of incorporating prior knowledge is the inclusion of more edges from the prior. In addition to this effect, due to the nature of lasso linear regression, addition of edges from the prior will yield information on the conditional independence between other edges. This could trigger the elimination of spurious edges in the estimated network, as we have seen in this application, where we found 4 edges inferred from the lasso method were not present in the network inferred by pLasso.

Here, we constructed two regulatory networks in different pathophenotypes, compared to a unified co-expression gene network in previous section. These two networks exhibit distinctive connection structures, including their topologies. In particular, the highly connected genes, or hub genes located at the functional center of the network, should have a high tendency to be associated with their phenotypic outcome. To verify this assumption, we selected the distinctive hub genes across different phenotype and then exploited their discriminating power of two disease states. We examined the 100 hub genes having most neighbors in the inferred network of patients showing distant metastasis, but not in the other group of metastasis-free patients. Among the 214 lymph-node-negative breast cancer patients we used to construct the gene networks, 107 showed evidence of distant metastasis and were considered as failure in our distant-metastasis-free survival analysis. For each of the hub genes we investigated, we divided the patients into two equal groups based on their expression values of the hub genes: the high-expression group and the low-expression group. We expected that for some of these hubs, the two groups would exhibit significant differences in their distant-metastasis-free survival outcome. To test this hypothesis, we used the "Survival" package in *R* to calculate the Kaplan–Meier survival curves. For each hub, its statistical significance was determined by controlling the false discovery rate at 0.2 with the Benjamini and Hochberg (BH) multiple testing procedure for the *p* values obtained from log-rank tests [31]. Based on this significance criterion, we found that the gene expression values of 18 % of hubs were significantly associated with breast cancer patient outcome. For the networks inferred from original lasso method, the expression values of 13 % of hubs showed significant association with breast cancer outcome. As the control, only 8 ± 2 % of 100 randomly selected genes demonstrated the significant association between their expression values and patient outcome. In addition, the list of selected gene hubs inferred from our pLasso method was used to predict survival outcome of an independent group of patients [32]. The identified gene signature achieved the highest sensitivity, with a comparable specificity among all the method we compared in the prognostic prediction results. These results together suggest we have successfully set up a framework capable of incorporating prior information for gene network reconstruction. As our knowledge of gene interactions accumulates over time, the pLasso method lends itself a potential improvement of performance since an increasing amount of prior knowledge gets incorporated into the analysis.

7.4 Identification of Protein Subnetworks as Biomarkers for Classifying Different Types of Brain Injuries

7.4.1 Neuroscience Research with Bioinformatics Approaches

Experimental advances in high-throughput technologies have provided neuroscientists a wealth of information that spans multiple levels of the nervous system. The rapid accumulation of genomic and proteomic information has provided valuable resources that motivate us to study how the genome as a whole contributes to the development, structure, and function of the nervous system. The intersection of bioinformatics and neuroscience research in the past 10 years has been focusing on the following areas: microarray, NGS, and proteomic techniques applied in brain diseases for disease biomarker identification; tools and methods in gene expression and sequencing study; methods for network analysis connecting molecular pathways to disease mechanisms and nervous function; and the development of user-friendly genome-scale resources such as the Allen Brain Atlas and BrainMap databases [33–35]. With large-scale genomic and proteomic information available, the omics or discovery-based approaches aim to extend the scope of brain research from individual genes or pathways to a system-level understanding of brain circuit function. These approaches have led to a revolution in the field of neuroscience research, by allowing high-throughput hypothesis generation, compared to the single hypothesis-testing approaches utilized in the traditional neurobiology laboratories. One area in which the omics-based approaches have been promising is in the detection of biomarkers for neurological and psychiatric diseases, as the focus is on generating novel hypotheses about gene signatures underlying disease mechanisms and discovering new therapeutic interventions. How to investigate brain functions in health and disease by integrating data simultaneously gathered at multiple levels of the nervous system remains a challenging task. Nevertheless, the strength of applying bioinformatics approaches combined with genomic and proteomic experimental techniques have been demonstrated in the area of feature gene identification. For example, a recent study by Emes et al. [36] combined genomic and bioinformatics approaches to successfully identify synaptic proteins that have changed during evolution and investigated how these proteins might relate to brain anatomy and function. Another study applied bioinformatics approaches to identify a unique gene signature that distinguishes familiar Alzheimer's disease mutation carriers from their normal siblings [37]. While the power of omics-based approaches has been clearly demonstrated in these studies, the adoption of these approaches has been challenging in the field of neuroscience. This is mainly due to the extreme heterogeneity and complexity of brain systems relative to other non-neural tissues. The complexity and the volume of the information available also have posed challenges in data sharing and modeling. New tools are required for encompassing multiple levels of the nervous system to allow us both to find patterns in the data and to test specific hypotheses. We expect the signaling or metabolic networks inferred by these tools help to guide the diagnosis and the treatment of neurological diseases. In Sects. 7.4.2 and 7.4.3, we have undertaken a systematic and quantitative study of TBI, a leading cause of death and disability in industrialized countries with approximately 1.7 million people sustaining a TBI in the USA each year (TBI statistics provided by the Centers for Disease Control).

7.4.2 Classifying Traumatic Brain Injuries with Network Biomarkers

TBI results from an external force causing immediate damage to brain tissue, followed by secondary pathogenic events which ultimately give rise to neurodegeneration. Dependent on the context of the primary injury, different cell responses are initiated that can exacerbate the injury to varying degrees. Cell death resulting from the initial impact on the brain tissue is irreversible so treatments normally focus on minimizing the secondary injury due to these cell responses. To date, these secondary injury responses have not been well characterized, leaving molecular classification of TBI cases difficult [38]. TBI remains a leading cause of death and disability in the industrialized countries and represents a growing health problem [39]. Thus, even a modest improvement in patient outcome could have significant public health benefits. An important topic in TBI study is to identify sensitive and specific biomarkers of TBI with diagnostic capabilities. Because for the mild TBI patients, 40–50 % suffer persistent neurological problems from one to three months following injury and 25 % after one year, it is often challenging to tell from traditional neurological scale whether the patient has experienced mild or moderate TBI. So effective biomarker identification can provide some insights on the early diagnosis and therapeutic development for treating mild or moderate TBI patients. TBI subtype classification is an important step toward the development and selective application of novel treatments.

Here, we aim to improve the identification of biomarkers that can distinguish the cortical contusion injury (CCI) and fluid percussion injury (FPI) in rodent animal models, representing focal and diffuse TBIs, respectively. If successfully identified, these biomarkers could be used to better direct treatments to TBI patients, and more optimistically, they could be potential targets of novel treatments. We have performed classification analyses of TBI cases using gene expression profiles. Typically, in expression profiling studies, genes that are not significantly differentially expressed between classes of genes (i.e., genes that are not associated with a class of TBI at a significance threshold) are neglected. These discarded genes' modest association may represent false negatives and may actually be important biomarkers of TBI. We hypothesize that these genes may be identified within functional units of genes that in aggregate have a significant association to a TBI class. To test this hypothesis, we identify biomarkers not as individual genes but as gene subnetworks by incorporating the gene expression profiles from CCI and FPI insults and the protein–protein interaction information from existing databases. The protein–protein interaction data provides direct information on specific protein relationships occurring along the backbone of the signaling network, while the whole-genome expression profiles are currently the largest source of high-throughput genomic information available, providing gene expression information on thousands of genes in different cells, tissues, or pathological specimens under various conditions. We expect the resulting subnetworks can provide novel hypotheses for testing the role(s) of pathways involved in different TBI classes.

7.4.3 *Overview of Subnetwork Biomarker Identification*

A binary protein interaction network was constructed from three sources of protein interactions, including the HPRD and BioGRID databases, and the interactions defined from an experiment by Chuang et al. [40], Peri et al. [41], Stark et al. [42]. This combined resource results in a protein interaction network with 8781 genes and 27,829 edges. Three types of brain injuries were applied to rats in a laboratory environment: CCI, FPI, and blast injury (as control). Gene expression values were then overlaid on the protein network. Each edge of this network was weighted by the level of co-expression between its two corresponding genes using Spearman

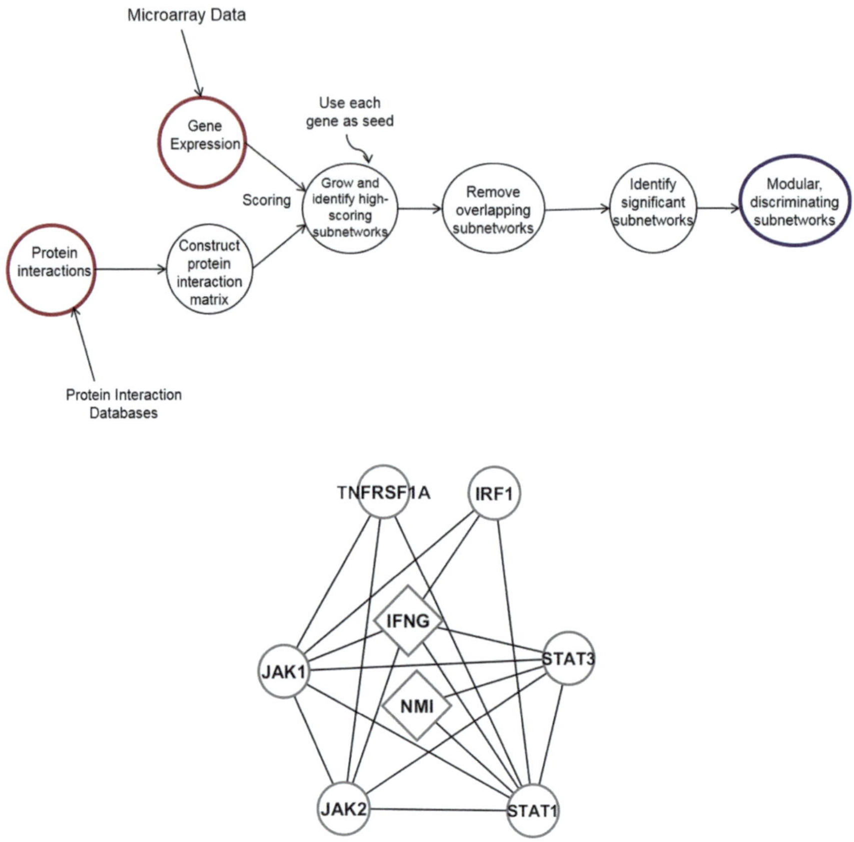

Fig. 7.6 Subnetwork identification for TBI classification. *Top panel* Overview of subnetwork identification approach. *Bottom panel* An example of discriminative subnetwork. The overall expression activity of the shown subnetwork is different between the mild cortical contusion injury (CCI) and the mild fluid percussion injury (FPI) samples. Nodes and edges represent proteins and protein interactions, respectively. The *circle-shaped* nodes indicate the corresponding genes are significantly up-regulated in FPI samples (FDR controlled q value <0.05), but not in CCI samples. The *diamond-shaped* nodes indicate the corresponding genes are not differentially expressed between FPI and CCI samples

correlation. Figure 7.6 demonstrates the overview of our subnetwork identification approach.

We define the subnetwork scoring function S, where M is defined as the modularity of the subnetwork and R is a measure of class relevance (or the discriminatory power of the subnetwork nodes to identify classes): $S = \beta_1 M + \beta_2 R$. Here, β_1 and β_2 allow us to weight the effects of the modularity and class relevance on the subnetwork score. To get a measure of how strongly the nodes within a module are connected, the modularity M is calculated as the mean of the clustering coefficients C_i for each nodes of the subnetwork, where C_i is defined as described in Dong and

Horvath [42]. *R* is a measure of the ability for a module to distinguish two classes. A *t* test is used to compare the subnetwork expression values for the samples of two classes, so $R = T(v_1, v_2)$, where an expression value v_j is simply an average of the normalized gene expression values (*z*-scores) for each node of the subnetwork for a given sample *j*.

To identify subnetworks that discriminate CCI, FPI, and controls, nodes were scored comparing each of these classes against the other two classes. In each of these settings, subnetworks were identified that discriminate a single class from the others. Individual differentially expressed genes were used as seeds for growing potential subnetworks. From these seeds, two neighboring nodes were iteratively added to the seed and subnetwork scores were recalculated. The pair of neighboring nodes that gave the biggest improvement in subnetwork score was added to the seed to form an initial subnetwork of three genes (i.e., an initial triangular subnetwork). Single neighbor nodes were then added iteratively until the subnetwork score could no longer be improved. We noted genes were shared across subnetworks, resulting in redundant subnetworks. We removed the low-scoring subnetwork if there is a significant overlap (>50 %) between a high-scoring subnetwork and itself. This process resulted in a set of subnetworks containing relatively unique genes, differentiating between CCI, FPI, and controls. To select the significant subnetworks, we performed two tests of significance. The first one generates the null distribution by permuting the expression vector of the gene in the network, so this dissociates the relationship between interaction and expression. The score of each identified protein subnetwork is indexed on the null distribution of all random network scores. Only those with *p* value smaller than 0.01 will be selected for further analysis. The second significance test generates the null distribution by permuting the phenotype, again only the networks with an empirical *p* value smaller than 0.01 will be selected.

Unlike traditional expression profiling methods, our network-based analysis can identify genes that are not differentially expressed and are often neglected. We determine whether such genes are essential for maintaining the integrity of a subnetwork whose overall expression is discriminative between samples. An example of the resulting discriminative subnetwork is shown in Fig. 7.6. The genes interferon-γ (IFNG) and myc-interactor (NMI) did not show significant differentiated expression between CCI and FPI samples, but they played an important role in the discriminative subnetwork by interconnecting many differentially expressed genes, such as JAK1, JAK2, JAK3, STAT1, and STAT3. Given the fact that both IFNG and NMI genes are well-known players in the cytokine signaling pathway involved in inflammatory response, our results suggest they can serve as potential targets for intervention. In addition, another advantage of our network-based analysis demonstrated from our preliminary study is that the list of identified significant gene subnetworks achieves higher sensitivity and specificity in classifying the heterogeneous responses corresponding to different classes of TBI, compared to a conventional analysis using an individual gene list. We therefore believe effectively incorporating gene expression profiles into protein interaction information can identify functional subnetworks that better classify different classed of TBI and are more reproducible across related studies than individual genes selected without network information. We understand that translating the knowledge gained from an animal model to molecular biomarkers identification in patients is practically challenging, simply because the brain tissue in TBI patients is rarely available, but the use of peripheral tissues such as lymphoblast or blood could be a potential solution. If successful, these identified biomarkers could be used to better direct the early diagnosis and treatment to TBI patients, and more optimistically, they could help to develop rationale-based therapies for treating the millions of Americans who suffer from TBI.

7.5 Conclusion and Future Directions

Feature selection is never a trivial problem, particularly in high-dimensional data analysis where few dozens of informative variables are often dispersed over a noisy background with thousands of non-informative variables. Network analysis plays an increasingly important role in the exploration of information communication and has been used to study the information on the relationship between genes or proteins. We envision that by leveraging the structure of gene co-expression network and protein interaction information, we successfully develop variable selection methods that aim to better identify feature genes associated with patient clinical parameters. Our work has set up a framework that can be easily generalized and extended to incorporate prior information of different types for feature gene identification.

Neuroscience has no doubt provided a rich application area for informaticians. More than a decade after the human genome sequencing was completed, there is a high demand for bioinformatics tools to explore a wealth of neuroscience information at multiple levels of nervous system, spanning from molecules to behavior. Our work provides an example on how bioinformatics approaches can be applied in neuroscience research by performing a genome-wide data analysis to gain a better understanding of interacting signaling pathways. The developed approaches will enable the generation of hypotheses subject to experimental validation. The resulting experimental data will, in turn, be used to generate more refined models that will advance our understanding of brain function.

References

1. Majewski IJ, Bernards R. Taming the dragon: genomic biomarkers to individualize the treatment of cancer. Nat Med. 2011;17(3):304–12.
2. Kohavi R, John G. Wrappers for feature subset selection. Artif Intell. 1997;97:52.
3. Yu L, Liu H. Efficient feature selection via analysis of relevance and redundancy. J Mach Learn Res. 2004;5:20.

4. Dy FG, Brodley CE. Feature selection for unsupervised learning. J Mach Learn Res. 2004;5:45.
5. Law MH, Jain AK, Figueiredo M. Feature selection in mixture-based clustering. In: NIPS; 2002. p. 8.
6. Alelyani S, Tang J, Liu H. Feature selection for clustering: review. In: Aggarwal C, Reddy C, editors. Data clustering: algorithms and applications. Boca Raton: CRC Press; 2013.
7. Cawley GC, Talbot NL, Girolami M. Sparse multinomial logistic regression via bayesian l1 regularisation. In: Neural information processing systems. 2006.
8. Mitra P, Murthy CA, Pal S. Unsupervised feature selection using feature similarity. IEEE Trans Pattern Anal Mach Intell. 2002;24:12.
9. He X, Cai D, Niyogi P. Laplacian score for feature selection. Adv Neural Info Process Syst. 2006;18:8.
10. Golub T, et al. Molecular classification of cancer: class discovery and class prediction by gene expression monitoring. Science. 1999;286(5439):531–7.
11. Wang Z, et al. Improving the sensitivity of sample clustering by leveraging gene co-expression networks in variable selection. BMC Bioinf. 2014;15(1):153.
12. Wang Z, et al. Spectral feature selection and its application in high dimensional gene expression studies. In: Proceedings of the 5th ACM conference on bioinformatics, computational biology, and health informatics. ACM; 2014.
13. Wang Z, et al. Incorporating prior knowledge into Gene network study. Bioinformatics. 2013;29(20):2633–40.
14. Wang Z, et al. A Bayesian framework to improve microRNA target prediction by incorporating external information. Cancer Info. 2014;13(Suppl 7):19.
15. Strehl A, Ghosh J. Cluster ensembles—A knowledge reuse framework for combining multiple partitions. J Mach Learn Res. 2003;3:35.
16. Zhang B, Horvath S. Stat Appl Genet Mol Biol. 2005;4 (Article17).
17. Qiu P, Gentles AJ, Plevritis SK. Discovering biological progression underlying microarray samples. PLoS Comput Biol. 2011;7(4):e1001123.
18. Witten D, Tibshirani R. A framework for feature selection in clustering. J Am Stat Assoc. 2010;105(490):14.
19. Golub TR, et al. Molecular classification of cancer: class discovery and class prediction by gene expression monitoring. Science. 1999;286(5439):531–7.
20. Alon U, et al. Broad patterns of gene expression revealed by clustering analysis of tumor and normal colon tissues probed by oligonucleotide arrays. Proc Natl Acad Sci USA. 1999;96 (12):6745–50.
21. Dudoit S, Fridlyand J. A prediction-based resampling method for estimating the number of clusters in a dataset. Genome Biol. 2002;3(7):RESEARCH0036.
22. Getz G, Levine E, Domany E. Coupled two-way clustering analysis of gene microarray data. Proc Natl Acad Sci USA. 2000;97(22):12079–84.
23. Meinshausen N, Buhlmann P. High dimensional graphs and variable selection with the lasso. Ann Stat. 2006;34:27.
24. Kramer N, Schafer J, Boulesteix AL. Regularized estimation of large-scale gene association networks using graphical Gaussian models. BMC Bioinform. 2009;10:384.
25. Parikh AP, et al. TREEGL: reverse engineering tree-evolving gene networks underlying developing biological lineages. Bioinformatics. 2011;27(13):i196–204.
26. Meinshausen N, Bühlmann P. High-dimensional graphs and variable selection with the lasso. Ann Stat. 2006;1436–1462.
27. Tibshirani, R. Regression shrinkage and selection via the lasso, J Royal Stat Soci Series B. 1996;58:22.
28. Wang Y, et al. Gene-expression profiles to predict distant metastasis of lymph-node-negative primary breast cancer. Lancet. 2005;365(9460):671–9.
29. Chen Y, Park B, Han K. Qualitative reasoning of dynamic gene regulatory interactions from gene expression data. BMC Genom. 2010;11(Suppl 4):S14.

30. Gusev Y, et al. In silico discovery of mitosis regulation networks associated with early distant metastases in estrogen receptor positive breast cancers. Cancer Inform. 2013;12:31–51.
31. Benjamini Y, Hochberg Y. Controlling the false discovery rate: a practical and powerful approach to multiple testing. JR Stat Soc. 1995;57(1):289–300.
32. van de Vijver MJ, et al. A gene-expression signature as a predictor of survival in breast cancer. N Engl J Med. 2002;347(25):1999–2009.
33. Bayes A, Grant SG. Neuroproteomics: understanding the molecular organization and complexity of the brain. Nat Rev Neurosci. 2009;10(9):635–46.
34. Laird AR, et al. ALE Meta-analysis workflows via the brainmap database: progress towards a probabilistic functional brain atlas. Front Neuroinform. 2009;3:23.
35. Zaldivar A, Krichmar JL. Allen Brain Atlas-driven visualizations: a web-based gene expression energy visualization tool. Front Neuroinform. 2014;8:51.
36. Emes RD, et al. Evolutionary expansion and anatomical specialization of synapse proteome complexity. Nat Neurosci. 2008;11(7):799–806.
37. Nagasaka Y, et al. A unique gene expression signature discriminates familial Alzheimer's disease mutation carriers from their wild-type siblings. Proc Natl Acad Sci USA. 2005;102 (41):14854–9.
38. Gaetz M. The neurophysiology of brain injury. Clin Neurophysiol. 2004;115(1):4–18.
39. Albert-Weissenberger C, Siren AL. Experimental traumatic brain injury. Exp Transl Stroke Med. 2010;2(1):16.
40. Chuang HY, et al. Network-based classification of breast cancer metastasis. Mol Syst Biol. 2007;3:140.
41. Peri S, et al. Development of human protein reference database as an initial platform for approaching systems biology in humans. Genome Res. 2003;13(10):2363–71.
42. Stark C, et al. BioGRID: a general repository for interaction datasets. Nucleic Acids Res. 2006;34(Database issue):D535–9.
43. Dong J, Horvath S. Understanding network concepts in modules. BMC Syst Biol. 2007;1:24.

Author Biographies

Dr. Zixing Wang got his Ph.D degree in genome science and technology at the University of Tennessee in 2011, with the thesis on the TGF-beta function in neuron development. Not totally fulfilled by his research experience in he field of experimental biology, he moved on and pursued his research interests in the field of bioinformatics and computational biology. Currently, he is working as a research fellow at University of Texas MD Anderson Cancer Center. His main research interest focuses on data mining and machine learning. He has been working on developing statistical methods to analyze large-scale genomic data and reconstruct gene regulatory networks, and applying the methods in the areas of cancer genomic and precision medicine. Dr. Wang has published 12 scientific papers in international well-recognized journals.

Dr. Wenlong Xu got his Ph.D degree in Biomedical Engineering at the University of Science and Technology of China in 2008, with the thesis on tumor subtypes classification based on gene expression. After working in the Data Center at the Baidu, Inc., the world renowned Chinese Web services company, he joined the University of Texas Health Science Center at Houston as a postdoctoral fellow in 2013. Dr. Xu has extensive experience in the areas of machine learning and pattern recognition. His current research focuses on microRNA target prediction, microRNA-mediated gene regulation, gene network reconstruction, and mobile health apps analysis. Dr. Xu has published 8 scientific papers in high-impact journals in the field of bioinformatics and mobile health.

Dr. Yin Liu is currently an associate professor in the Department of Neurobiology and Anatomy at the University of Texas Health Science Center at Houston, and an adjunct faculty member in the Department of Biomedical Engineering at the University of Texas at Austin. She received her Ph.D in Computational Biology and Bioinformatics from Yale University in 2007. Subsequently, she joined the University of Texas Medical School at Houston and established her own research group. Within the domain of bioinformatics, her laboratory develops computational and statistical methods to analyze and integrate heterogeneous data sources for understanding how biological interactions modulate complex signaling responses. In the past few years, she has built on her knowledge of gene networks and began to explore another fascinating problem in the post-genomic era, that is, to identify gene signatures and pathways that are involved in common cancers and neurological diseases from large-scale genomic data analysis. Dr. Liu has published more than 20 scientific papers in internationally well-recognized journals and has written three book chapters. She has an h-index of 11 and a total of more than 700 independent citations. Dr. Liu has frequently participated in grant reviews, including NIH, NSF, and the private foundations. She is on the editorial board of the Journal, *Frontiers in Statistical Genetics and Methodology*. She has served as the program committee members for two scientific conferences and is a frequent journal reviewer for many top-tier journals in the field of bioinformatics and systems biology. For more information, please visit: http://nba.uth.tmc.edu/homepage/liu/.

MIX
Papier aus verantwortungsvollen Quellen
Paper from responsible sources
FSC® C105338
FSC
www.fsc.org

If you have any concerns about our products,
you can contact us on
ProductSafety@springernature.com

In case Publisher is established outside the EU,
the EU authorized representative is:
Springer Nature Customer Service Center GmbH
Europaplatz 3, 69115 Heidelberg, Germany

Printed by Libri Plureos GmbH
in Hamburg, Germany